A Doctorate and Beyond

Graham C. Goodwin · Stefan F. Graebe

A Doctorate and Beyond

Building a Career in Engineering
and the Physical Sciences

Illustrations by Adam Sandford

Graham C. Goodwin
Centre for Complex Dynamic Systems
  and Control
University of Newcastle
Callaghan, NSW
Australia

Stefan F. Graebe
MyWings Coaching Hout Bay
Cape Town
South Africa

ISBN 978-3-319-45876-2   ISBN 978-3-319-45877-9 (eBook)
DOI 10.1007/978-3-319-45877-9

Library of Congress Control Number: 2016950881

© Springer International Publishing AG 2017
This work is subject to copyright. All rights are reserved by the Publisher, whether the whole or part of the material is concerned, specifically the rights of translation, reprinting, reuse of illustrations, recitation, broadcasting, reproduction on microfilms or in any other physical way, and transmission or information storage and retrieval, electronic adaptation, computer software, or by similar or dissimilar methodology now known or hereafter developed.
The use of general descriptive names, registered names, trademarks, service marks, etc. in this publication does not imply, even in the absence of a specific statement, that such names are exempt from the relevant protective laws and regulations and therefore free for general use.
The publisher, the authors and the editors are safe to assume that the advice and information in this book are believed to be true and accurate at the date of publication. Neither the publisher nor the authors or the editors give a warranty, express or implied, with respect to the material contained herein or for any errors or omissions that may have been made.

Printed on acid-free paper

This Springer imprint is published by Springer Nature
The registered company is Springer International Publishing AG
The registered company address is: Gewerbestrasse 11, 6330 Cham, Switzerland

Preface

The book addresses two groups of readers

- A specific group, comprising anybody contemplating, or already engaged in, a doctorate in engineering or one of the physical sciences
- A wider group, comprising anybody who is interested in topics such as the ethics of research, student–supervisor interactions, managing success, setting goals, achieving an appropriate work–life balance or recovering from injustice.

Parts I and II cover topics associated with doing a doctorate. Although the material is principally directed at students in engineering or the physical sciences, many of the ideas apply more broadly to doctoral students in any faculty or discipline. The material may also interest teachers, counsellors and parents of young people contemplating this career choice.

Parts III and IV focus on issues that you will face as your career progresses into ever more responsible positions. Much of the material applies to a wide group of readers. In particular, Part IV is aimed at anybody moving into a senior position whether or not they hold a doctoral qualification.

This is not a technical book about your field of study. Rather, it is a resource to motivate and galvanize your decision-making as you manage your studies and subsequently build your career.

The book aims to support you all the way from your graduate studies, when you contemplate whether to pursue a doctorate, through to completing your degree and your career beyond. By "beyond" we refer to your postgraduate career from junior to more senior stages, whether you devote them to the academic, private or government sectors of society.

We strive to inspire your decision-making by both highlighting important questions and by providing guidance in finding your answers. We deliberately call them "your" answers rather than simply "answers" because there are no ready-made answers that suit everybody, particularly not in a rapidly changing world.

The book is structured into the following four parts.

Fig. 1 "Research can be an ultimate expression of 'self', similar to painting or performing and composing music"—see Sect. 1.2.3

Part I: Contemplating Whether to Pursue a Doctorate

We believe that becoming aware of and contemplating questions such as:

- Should you do a doctorate?
- How do you choose where to do your doctorate?
- How do you choose a supervisor?
- How and where could you use your doctorate if you obtain one?

Will aid your decision whether or not to enrol for a doctorate. It will also help you consider options that will broaden your spectrum of choice.

Part II: During Your Doctorate Studies

Questions that seek to help and guide you while writing a doctoral thesis are addressed in Part II. They include:

- What outcomes will be expected?
- How to establish good student–supervisor interactions?
- What are the benefits of networking?
- What tools are needed to undertake doctoral studies?
- What ethical issues arise?
- How do you write a doctoral thesis?

Part III: Once You Have a Doctorate and Are Embarking on Your Early Career

This group of readers is addressed in Part III of the book. Here we inspire by reflecting on questions such as:

- What career opportunities exist in the academic and private sectors?
- How does one move between sectors?
- What is the difference between administration, management, leadership and entrepreneurship?
- How does one mentor young colleagues?
- How does one obtain funding?
- How does one kick start a cycle of success?

Part IV: Once You Have Been Working for Some Time and are About to Move into a More Senior Role

Part IV of the book is concerned with more senior career phases. Most of this material applies more broadly than is suggested by the title of the book and is relevant whether or not you hold a doctorate.

- How to continue growing your cycle of success?
- How to deliver a great public speech?
- How to prepare for, and recover from, injustice?
- How to deal with the media?

- How to achieve a work–life balance that is right for you?
- How to keep the bigger picture in mind?

These questions, and more, are addressed in Part IV.

When to Read the Different Parts of the Book?

Whilst each part of the book addresses a different phase of your career, we strongly advocate that you visit different parts of the book on multiple occasions. Reading later parts will inform you of broader issues and provide stretch information.[1] Reading earlier parts will remind you how to advise younger people who are passing through that stage.

Each chapter contains an overview and a comprehensive summary that are meant to aid previewing "stretch information" or as a quick review of earlier career phases when you are mentoring junior colleagues.

A Changing World

We live in a rapidly changing world. One thing is for certain: there will be significant changes during the time span of your career. Therefore this guide focuses on principles that will remain relevant during the next decades of your life. As we discuss the principles, we also add specifics and examples to make our points more tangible. However, as you consult the book, try to focus on the more timeless principles and adapt the specifics to emerging technologies and contemporary best practice during your career.

We quickly illustrate with three examples.

1. Networking and reputation: The means and technologies you use to network will keep changing; but the importance of networking will always remain high. Similarly, aspects that build a good reputation may change—but having a good reputation will always remain crucial. Reputations are hard to build and easy to destroy.
2. Technologies and required skills: Of course these will change and evolve. However, their central importance will remain. The book is not intended to teach the current state of the art of technology. Indeed, if it did so, it would quickly become obsolete. Rather we point to the role technology takes as you manage your studies and career.

---

[1] "Stretch information," i.e. information that stretches beyond the reader's present situation.

3. Leadership paradigms: These will also shift over time. Thus we will concern ourselves less with covering details than we will with conveying how being conscious and mindful of your leadership style is a core success factor.

While the specifics of these three principles will clearly change over time, their importance will remain. It is always true that poor leadership, mediocre skills or a destroyed reputation will work against you. Yet, if you look around you, or read the news, you will see that these errors occur all the time. On the other hand, you will observe that successful people continuously hone their leadership style, keep their skills up to date and fervently protect their reputation.

We believe that becoming conscious and mindful of these principles is of paramount importance, and we believe that this book will help you achieve these goals.

Sources

In writing the book we have mainly drawn upon our personal experiences. These experiences were shaped, shared and created by a large number of colleagues, students, mentors and professionals drawn from a broad spectrum of society including academia and the private sector.

We quote many of these experiences directly. Others are reflected indirectly by virtue of the fact that they have left a mark on us and influenced our views.

Since it is personal interactions that we have benefitted most from, we have chosen an equally personal and anecdotal tone for the book rather than aiming for a research-orientted style.

Consequently, we have abstained from scientific style referencing. We do offer selected sources that you may care to consult for further reading. However, neither the style nor scope are as comprehensive as they would be in a research treatise. We address topics as wide-ranging as leadership, job interviewing techniques, ethics, work–life balance, writing grant applications and delivering conference presentations. Volumes have been written about each of these topics in their own right.

All of the above topics will play crucial roles during the various phases of your career. The objective of the book is to bring all of the core ideas together in one place and to put them into perspective. To learn more about the specifics and current "state-of-the-art" thinking in any of the component topics, we refer you to contemporary sources.

Book Website

There is also a book website: www.ADoctorateAndBeyond.com.

This website is under continuous development and will provide up-to-date information on certain topics of relevance to the book.

Times of "Luck", Times of "Hardship"

All that we have mentioned so far refers to your active role in pursuing your degree and your subsequent career. However, in life, there are always forces beyond our control, some work for us, some against us. For the sake of simplicity we refer to the former simply as "luck", the latter as "hardship". In reality, the complexities of life are rarely that cleanly cut. However, the book will support you during either of these times.

If you talk to successful people in any field you will find that some have, more or less, stumbled into their careers, taking opportunities that opened up along the way and discovering skills and talents as they went along. Others chose careers as they followed someone else. Others have started out wanting to prove that "they can do it".

To what extent you "plan" and to what extent you "leave things to chance" is a matter of personal preference and can change depending upon the circumstances. We aim to support you both during times when you choose to plan and times when things evolve by chance. Experience suggests that you will always encounter coincidences and unforeseen opportunities along the way.

Times of planning, times of luck, times of hardship: we hope to provide you with a resource of inspiration.

Callaghan, Australia Graham C. Goodwin
Cape Town, South Africa Stefan F. Graebe

Acknowledgments

The seeds of this book were sown when Graham visited Universidad Técnica Federico Santa María in Valparaíso, Chile approximately 20 years ago. Graham's host during that visit was Mario Salgado. He asked Graham to give a different kind of seminar with the intriguing title "The responsibilities of Ph.D. students and their supervisors;" a tough call. Over the years, this seminar evolved and Graham presented it to many different groups. However, a book was far from Graham's mind. Then in 2010, a dear friend of Graham's who worked in industry, John Edwards, said he would like to meet Graham to discuss career and life issues, contrasting and comparing industry and academia. At the time, John was very ill and so, in the spirit of the well-known book by Mitch Alborn, "Tuesdays with Morrie", it became "Fridays with John". A book project was discussed. Indeed a rough draft was written but it was never finished. By now, sufficient momentum had been achieved that a book became a real possibility. Finally, Graham approached Stefan. Graham had worked with Stefan previously and knew he had considerable experience using a doctorate in industry. This nicely complemented Graham's own, more academically orientated, background. So the book became a reality.

During initial stages, the authors discussed the book with many colleagues and friends. Almost all said one of two things, namely, "I definitely want to read this book" or "I wish I had access to this book 10 years ago".

In writing the book, the authors received advice and help from many former students, colleagues and friends. Some of these are mentioned below. However, in truth, the book brings together ideas and experiences gathered over a lifetime from all of the author's colleagues.

The following people made specific contributions to the book by reading early drafts and often by writing a paragraph or two on an issue that affected them directly.

- Brian Anderson (Australian National University, Australia)
- Diego Carrasco (Research Academic, Australia)
- Roger Davies (Senior Industrial Manager, UK and Australia)
- Ramon Delgado (Research Academic, Australia)

- Barbara Elion (One Life Media, South Africa)
- Niki Enescu (Petrom, Romania)
- Arie Feuer (Technion, Israel)
- Jan Graebe (University of Göttingen, Germany)
- Chomba Hermanus (MLDesign, South Africa)
- Sarah Johnson (Mid Career Academic, Australia)
- Max Linskeseder (OMV, Austria)
- Wendy Mason (Chief Education Officer, Australia)
- Brian May (Astrophysicist and Musician, UK)
- Tamas Mayer (OMV, Turkey)
- Richard Middleton (University of Newcastle, Australia)
- Steve Mitchell (AMP Control, Australia)
- Raheleh Nazari (Doctoral Student, Australia)
- Barbara Oberhauser (OMV, Austria)
- Turker Ozkocak (Shell, USA)
- Alan Roberts (University of Newcastle, Australia)
- Mario Salgado (UTSFM, Chile)
- Thomas Schön (Uppsala University, Sweden)
- Titilayo Seriki (Cielarko, South Africa)
- Meng Wang (Ericsson AB, Sweden)
- Torjörn Wigren (Ericsson AB and Uppsala University, Sweden)
- Adrian Wills (University of Newcastle, Australia)

Importantly, we wish to pay tribute to our Editor, Oliver Jackson, who had the courage to take on this different book project. He was interested and supportive from the first meeting. He obtained feedback from six anonymous reviewers. These reviews were extremely helpful and contained many great suggestions. All of these have been addressed, in one form or another, in the final version.

We wish to thank Jayne Disney who not only did all the typing and layout of the book, but also lifted our spirits when they sagged during the writing process.

And, on a personal note, we thank our partners Rosslyn Goodwin and Tania Steincke for support on the work side of our work–life balance and for love on the other side.

Contents

Part I Choosing Whether or Not to Do a Doctorate

1 Is a Doctorate the Right Course for You? 3
   1.1   Overview 3
   1.2   Reasons for Doing a Doctorate........................ 4
         1.2.1   You Are Planning a Career that Requires
                a Doctorate Qualification...................... 4
         1.2.2   You Are Fascinated by One or More Fields
                in Engineering or Science 4
         1.2.3   You Are Interested in Science, Technology
                or Mathematics Just for the Sake of It............ 5
         1.2.4   You Want to Prove to Yourself that You Can
                Achieve the Highest Possible Academic
                Qualification 5
   1.3   A Master's Degree: A Possible Milestone on the Way....... 6
   1.4   Summary 7
   1.5   Further Reading 7

2 How Might You Use Your Doctorate, if You Get One?.......... 9
   2.1   Overview 9
   2.2   Positions that May Require a Doctorate 9
         2.2.1   University Employment........................ 9
         2.2.2   Private Sector 10
         2.2.3   Government Sector 11
         2.2.4   A Private Consultant 11
   2.3   Skills and Passion.................................... 12
   2.4   Summary 13
   2.5   Further Reading 13

3	The Where, Who and What of Doing a Doctorate		15
	3.1	Overview	15
	3.2	Selecting an Institution	15
		3.2.1 What to Look for?	15
		3.2.2 How to Check These Attributes	16
	3.3	Selecting a Supervisor	18
	3.4	Selecting a Topic	20
	3.5	Summary	23
	3.6	Further Reading	23
4	How Hard Will You Have to Work?		25
	4.1	Overview	25
	4.2	An All Consuming Activity	25
	4.3	Things to Avoid	25
	4.4	Progress	26
	4.5	Impact on Your Family	27
	4.6	Summary	28
	4.7	Further Reading	28
5	How Long Will It Take You to Get a Doctorate?		29
	5.1	Overview	29
	5.2	Mission Creep	29
	5.3	Be Focused	30
	5.4	Goal Setting	31
	5.5	Can, and Should, You Do a Doctorate Part-Time?	32
		5.5.1 Balancing Work and Thesis	32
		5.5.2 Making a Success of a Part-Time Doctorate	33
	5.6	Summary	35
	5.7	Further Reading	35

Summary of Part I

Part II Doing Your Doctorate

6	How to Begin		41
	6.1	Overview	41
	6.2	A Guiding Principle	41
	6.3	Hit the Road Running	41
	6.4	Keeping Records	42
	6.5	Working in a Group	42
	6.6	Working with Industry	43
	6.7	Summary	44
7	Student–Supervisor Interactions		47
	7.1	Overview	47
	7.2	Engagement	47

	7.3	Can I Work with Others as Well as My Supervisor?	49
	7.4	Can and Should I Spend Time at Another Institution?	50
	7.5	Swapping Topic	51
	7.6	Changing Supervisor	52
	7.7	Summary	53
	7.8	Further Reading	54
8	**The Value of Networking**		**55**
	8.1	Overview	55
	8.2	Networking with Peers	55
	8.3	Other Networks	55
	8.4	Attending Conferences	56
	8.5	Joining Professional Associations	57
	8.6	Social Media	57
	8.7	Summary	58
	8.8	Further Reading	59
9	**Tools of the Trade**		**61**
	9.1	Overview	61
	9.2	Have Courage and Tenacity	61
	9.3	Setting up an Experimental Facility	62
	9.4	Keeping Track of the Literature	62
		9.4.1 How Much Time Should I Spend Reading the Literature?	62
		9.4.2 Keeping Track of References	63
	9.5	Other Tools	64
	9.6	Work-Life Balance	64
	9.7	A Long-Term Vision	65
	9.8	Summary	66
	9.9	Further Reading	67
10	**The Art of Publication**		**69**
	10.1	Overview	69
	10.2	Publications	69
	10.3	How Many Papers Should I Write?	70
		10.3.1 Conference and Journal Papers	70
		10.3.2 Is It Possible that a Book Arises from My Thesis?	70
	10.4	Writing a Good Paper	71
	10.5	Peer Review and Resilience	72
	10.6	Writing Patents	74
	10.7	Summary	75
	10.8	Further Reading	76

11 The Art of Making Great Presentations ... 77
- 11.1 Overview ... 77
- 11.2 Conference Presentations ... 77
- 11.3 Three Minutes Is All You Have ... 79
- 11.4 Other Ideas ... 80
- 11.5 Summary ... 80
- 11.6 Further Reading ... 81

12 The Ethics of Research ... 83
- 12.1 Overview ... 83
- 12.2 The Context ... 83
- 12.3 Humans, Animals and the Environment ... 84
- 12.4 Keeping Accurate Records ... 84
- 12.5 Accuracy of Publications ... 84
- 12.6 Plagiarism ... 85
  - 12.6.1 Citing the Work of Others ... 85
  - 12.6.2 Citing Your Own Work ... 85
- 12.7 Multiple Submission of Papers ... 86
- 12.8 Who Should be an Author of Publications? ... 86
- 12.9 Summary ... 87

13 How to Write Your Thesis ... 89
- 13.1 Overview ... 89
- 13.2 When to Start Writing ... 89
- 13.3 Planning ... 90
- 13.4 Reviewing ... 90
- 13.5 The Target Length ... 91
- 13.6 Be Self-Critical ... 91
- 13.7 The Alternative Route of Presenting Papers Only ... 91
- 13.8 Summary ... 91
- 13.9 Further Reading ... 92

Summary of Part II

Part III   Using Your Doctorate: The Early Years

14 Securing a Job ... 97
- 14.1 Overview ... 97
- 14.2 Where to Apply ... 97
- 14.3 Selecting an Appropriate Level ... 98
- 14.4 Writing Successful Job Applications ... 99
- 14.5 Referees ... 100
- 14.6 The Job Interview ... 100
- 14.7 Summary ... 103
- 14.8 Further Reading ... 104

15 An Academic Position ... 105
- 15.1 Overview ... 105
- 15.2 Teaching ... 106
  - 15.2.1 Prepare Well ... 106
  - 15.2.2 Good and Bad Teaching ... 107
  - 15.2.3 Focusing the Content ... 108
  - 15.2.4 Communicating with Clarity ... 108
  - 15.2.5 Inspire ... 109
  - 15.2.6 Learn the Lessons of Failure ... 109
- 15.3 Research ... 110
  - 15.3.1 Good and Bad Research ... 111
  - 15.3.2 How to Get Started ... 111
  - 15.3.3 Attracting Research Funding ... 112
  - 15.3.4 Carrying Out the Research ... 113
  - 15.3.5 Time Management ... 114
  - 15.3.6 Reviewing Papers ... 114
  - 15.3.7 Building Momentum ... 115
  - 15.3.8 Multidisciplinary Research ... 116
  - 15.3.9 Making a Quantum Leap Forward ... 117
- 15.4 Publication ... 119
- 15.5 Administration ... 119
- 15.6 Community Engagement ... 120
- 15.7 Self-Reliance ... 120
- 15.8 Working with Industry ... 121
- 15.9 Summary ... 122
- 15.10 Further Reading ... 123

16 A Position in Industry ... 125
- 16.1 Overview ... 125
- 16.2 A Doctorate in Industry ... 125
- 16.3 Having and Delivering on Targets ... 126
- 16.4 S.M.A.R.T. Goals ... 127
- 16.5 From Administration to Entrepreneurship ... 129
  - 16.5.1 Administration ... 129
  - 16.5.2 Project Management ... 130
  - 16.5.3 Leadership ... 130
  - 16.5.4 Entrepreneurship ... 131
  - 16.5.5 An Illustration ... 132
- 16.6 Community Engagement ... 133
- 16.7 Summary ... 133
- 16.8 Further Reading ... 134

17 Moving Freely Between Academia and Industry ... 135
- 17.1 Overview ... 135
- 17.2 Moving Freely from Academia to Industry ... 135

	17.3	Moving Freely from Industry to Academia	136
	17.4	A Case Study of a Person Who Successfully Made the Transition in both Directions	137
	17.5	Holding a Joint Appointment Between Academia and Industry	139
	17.6	Summary	139
18	**The Cycle of Success**		**141**
	18.1	Overview	141
	18.2	Position Plus Reputation	142
	18.3	Skills Plus Resources	144
		18.3.1 Making a Spontaneous Pitch	144
		18.3.2 Making a Prepared Pitch	145
		18.3.3 Attracting the Right People	147
	18.4	Leadership	147
		18.4.1 Team Development	148
		18.4.2 Dealing with Mistakes	148
		18.4.3 Mentoring	149
		18.4.4 Friends and Enemies	150
		18.4.5 Dealing with Rejection	150
		18.4.6 Lead by Example Not by Brute Force	151
	18.5	Summary	151

Summary of Part III

Part IV Using Your Doctorate: The Later Years

19	**The Cycle of Success: The Later Years**		**157**
	19.1	Overview	157
	19.2	Position Plus Reputation	157
	19.3	Resources Plus Skills	158
	19.4	Delivering on Promises	159
	19.5	Efficient Conduct of Meetings	159
	19.6	Leadership	160
		19.6.1 Always Appoint the Very Best People	160
		19.6.2 Have Clearly Articulated Goals	160
		19.6.3 Believe in What You Are Doing	161
		19.6.4 Set Priorities	161
		19.6.5 Listen to Others	162
		19.6.6 Recognize Your Own Weakness	162
		19.6.7 Delegate	162
		19.6.8 Understand Teamwork	163
		19.6.9 Inspire	163
	19.7	Summary of Leadership Qualities	163
	19.8	Learning from Other Leaders	164

19.9	An Example of Great Leadership	164
19.10	Judging Others	164
19.11	Summary	165
19.12	Further Reading	166

20 Public Speaking and Dealing with the Media … 167
- 20.1 Overview … 167
- 20.2 Public Speaking … 167
- 20.3 Dealing with the Media … 169
- 20.4 Summary … 170

21 Job and Career Changes … 171
- 21.1 Overview … 171
- 21.2 Embracing Unsolicited Change … 171
- 21.3 Should You Remain Technically Active or Move into Management? … 172
- 21.4 Continuity and Leaps into the Unknown … 173
- 21.5 Summary … 174
- 21.6 Further Reading … 174

22 Mentoring and Succession Planning … 175
- 22.1 Overview … 175
- 22.2 Mentoring … 175
- 22.3 Succession Planning … 176
- 22.4 Retirement … 177
- 22.5 Summary … 177
- 22.6 Further Reading … 178

23 Work-Life Balance … 179
- 23.1 Overview … 179
- 23.2 Reflecting on Work-Life Balance … 179
- 23.3 Reacting Positively to a Forced Time Out … 181
- 23.4 Working Smarter Not Harder … 181
  - 23.4.1 Reduce Task Swapping … 182
  - 23.4.2 Task Completion … 182
- 23.5 Energy Management … 183
- 23.6 Delegating … 184
- 23.7 Summary … 184
- 23.8 Further Reading … 185

24 Keeping the Bigger Picture in Focus … 187
- 24.1 Overview … 187
- 24.2 Keeping the Bigger Picture in Focus … 187
- 24.3 Regaining Balance When Injustice Strikes … 189
- 24.4 Broader Life Issues … 190
- 24.5 Embracing Serendipity … 190

24.6	Our Final Wish for You	191
24.7	Summary	191
24.8	Further Reading	192

Summary of Part IV

About the Authors

Graham represents the experience of a career principally dedicated to academia. He holds a doctorate in engineering. He has been working in universities for over 40 years, has taught hundreds of students and supervised 40 doctoral students. He has also written 10 books, hundreds of publications and holds many patents. During his career he had served in numerous positions ranging from a junior academic to Dean of a Faculty. His research has attracted substantial external funding. He has been Director of several special research centres that he created. He was a board member of two high technology companies and Chairman of a university spin-off company. He is a Fellow of the Royal Society of London and a Foreign Member of the Royal Swedish Academy of Sciences.

Stefan represents the experience of using a doctorate in both university environment and in the private sector. He holds a doctorate in engineering. After some 40 publications and a book as Associate Professor, his doctorate was pivotal for him to move into the private sector as a control engineering expert in the oil industry. There he successfully translated his technical expertise into business and financial leadership. Within this context, he advanced through various executive positions including serving as chief strategist. He rose to managing director and held board director positions. He has managed companies with hundreds of millions of Euros annual turnover. In a third phase of his career he used his academic and business experience, together with additional training, to become a coach for individuals and businesses.

Graham is based in Newcastle, Australia; Stefan is based in Cape Town, South Africa. Both work globally.

Part I
Choosing Whether or Not to Do a Doctorate

Overview of Part I

In this part we discuss whether doing a doctorate is the right course of action for you. The first chapter guides you towards an initial answer by suggesting situations in which doing a doctorate would be an ideal choice for you.

The second chapter describes career options for doctoral graduates. You can assess whether any of these capture your fancy.

Subsequent chapters discuss issues such as how to find a supervisor, how to choose an institution and how to select a topic. We also discuss how hard you will have to work and how long it may take you, as well as the desirability of completing a doctorate part time.

This knowledge will then form a solid foundation upon which you can make your final decision as to whether or not pursuing a doctorate is the right course of action for you.

Chapter 1
Is a Doctorate the Right Course for You?

1.1 Overview

Doing a doctorate can be one to the most exciting things you will do in your life. This is a chance to do something that is, at the same time, intellectually challenging and massively rewarding. Having a doctorate has the power to create unparalleled life and work opportunities. As one of Australia's most senior academics (Professor Brian Anderson FRS from the Australian National University) recently said to Graham

> "Most people work to live but researchers typically live to work".

Moreover, doing a doctorate can dramatically change your self image. For example, Raheleh Nazari, who recently completed her doctorate in engineering, said "The doctoral program was a total life changing experience. It changed my complete personality. I became wiser, more patient and more able to focus on the important things in life. I now have a completely different image of myself and my capacity to contribute".

If you seek further affirmation of the capacity of research to be a positive force in your life, then you may care to quickly read Sect. 22.4 of the book.

Of course, doing a doctorate is an important decision. Thus it helps to have full knowledge of the steps needed to get a doctorate and the doors that will open if you have a doctorate.

In this context, this chapter is intended to give you guidance during your decision-making process. By the end of this chapter you should have arrived at a tentative indication of whether a doctorate is the best course of action for you or whether you might be better served by creating a career without a doctorate. Either of these can be a great choice depending upon your individual circumstances. Subsequent chapters will substantiate your initial decision by providing information about the process of doing a doctorate and career opportunities.

1.2 Reasons for Doing a Doctorate

There are four principal reasons that, should any of them apply to you, would suggest that a doctorate is the right course of action for you. These reasons are discussed below.

1.2.1 You Are Planning a Career that Requires a Doctorate Qualification

There are many jobs for which a doctorate is a necessary qualification. Examples in the private sector include employment in the research and development wing of a large corporation or becoming a high-level advisor. Examples in the government sector include tertiary education positions or government research organizations. If your interests lie in these areas, then you would pursue a doctorate as a necessary job requirement. To confirm your decision, it is a good idea to check job announcements in your area of interest or talk to friends who already have the kind of position in which you are interested.

The following comment was made by Dr. Torbjörn Wigren (Torbjörn is a senior engineer at Ericsson AB in Sweden. He also holds an Adjunct Professorship at Uppsala University).

> "My career goal was to obtain a challenging engineering position that focused on design and innovation rather than on simple implementation of the ideas of others. To achieve this in Sweden I knew that I needed to have a doctorate so I went about obtaining one".

1.2.2 You Are Fascinated by One or More Fields in Engineering or Science

If you are fascinated by exciting developments such as mobile telecommunications or biomedical engineering and you want to participate in the associated innovation process, then a doctorate may be the right course for you. For example, we know a student who was fascinated with sustainable energy vehicles and this motivated him to do a doctorate in a closely aligned field. Actually, this particular student grew up in Australia and did his honours degree at the University of Melbourne but he decided that the best place for him to pursue his specific dream was in Germany. So he left Australia to do his doctoral work overseas.

1.2 Reasons for Doing a Doctorate

> You will know that you fall into this category if a particular area of science or technology captures your imagination and interest.

Examples could arise across a broad spectrum including topics such as transportation, telecommunications, computing, medical instrumentation, geoscience; indeed any area of science or technology.

1.2.3 You Are Interested in Science, Technology or Mathematics Just for the Sake of It

You will know if you belong to this category if, by your inner nature, you always seek to understand science at a deep level. As one of our doctoral students recently said, "I was totally fascinated by the topic and was driven to delve more deeply into it". People in this category typically enjoy the sheer elegance of results, love reading popular books on science and enjoy the beauty of mathematical proofs.

For people in this category,

> Research can be an ultimate expression of "self", similar to painting or performing and composing music.

If this applies to you, then we recommend that you have the courage to pursue your dream without necessarily worrying about the immediate applicability of the work.

1.2.4 You Want to Prove to Yourself that You Can Achieve the Highest Possible Academic Qualification

Many people enjoy studying and are highly successful at doing it. They may thus feel that the progression on to a doctorate is a natural course for them. This can also be aided and supported by family encouragement.

As an example, we quote the experience of Dr. Roger Davies. Roger obtained a doctorate in Chemical Engineering from the University of Swansea, in the UK. His first job was in the research section of a Chemical company. From there he moved into senior management roles in Chemical Manufacturing. Roger said "My father had a tough upbringing and had a steely determination that all his children would receive the educational opportunities that he had not received. I was so influenced by my father's zeal that I wanted to 'achieve as much as I could academically'. This translated, in my mind, to achieving a doctoral degree. So I did - and he was so proud".

Roger added, "It is likely that many people will pursue a doctorate to prove to themselves, or others, that they can do it. For such people it is also worth focusing on future career options as soon as possible since this may steer the direction of the doctoral studies".

1.3 A Master's Degree: A Possible Milestone on the Way

For completeness, we also mention that many universities offer, or insist upon, a master's degree (M.A. or M.Sc.) as an intermediary degree between a bachelors and a doctorate degree. This can be a useful mechanism to see if you should continue with a doctorate.

The second author of this book, Stefan, took this option and completed a M.Sc. at the University of California, San Diego, before earning his doctorate at the Royal Institute of Technology in Stockholm, Sweden. At the time of completing his bachelor's degree, he was not yet one hundred percent sure if he wanted to specialize in control engineering. He therefore opted for an interdisciplinary project which would earn him a master's degree. It was this work[1] that firmly ignited his fascination with control engineering and ultimately led to him enrolling in a doctorate programme.

Here are some considerations that may encourage you to first opt for a master's degree before considering a doctorate:

- You know that you want to pursue further academic work after completing a bachelor's degree, but you are not (yet) sure if you want to go the whole way and pursue a doctorate.
- You want to gain further exposure to a specialty topic in order to then decide whether it ignites your passion sufficiently to go the next step to doctoral studies.
- It may be unclear whether a particular project has sufficient potential for a doctorate degree. In that case, you might "test" the topic in the master's environment. You might then consult your supervisor as to whether the topic could be extended to doctorate level.
- You are doing an industry-funded degree whilst working for a company (see also Sect. 5.5) and your company is only prepared to fund a master's degree. You can then try to convince them later to extend the funding for you to complete a doctorate by delivering outstanding master's work.
- Finally, there can be a strong practical reason; some doctorate programmes only accept students who already hold a master's degree.

The potential disadvantages include

- It might take you longer to get a doctorate if you pursue a master's degree along the way. This, however, also depends on how well your master's work is aligned with your subsequent doctoral work.
- It may be more difficult to obtain a scholarship for a master's than a doctoral degree.

---

[1] More will be said about this interdisciplinary project in Sect. 7.3.

1.4 Summary

The key thing to decide based on this chapter is whether you fall into one of the following four categories:

- You are planning a career that requires a doctorate qualification (see Sect. 1.2.1).
- You are fascinated by one or more fields in engineering or science (see Sect. 1.2.2).
- You are interested in science, technology or mathematics for the sake of it (see Sect. 1.2.3).
- You want to prove to yourself that you can achieve the highest possible academic qualification (see Sect. 1.2.4).

These are all driving forces that are typical motivations for students to pursue a doctoral degree.

You will be able to clarify your initial decision by reading the subsequent chapters which deal with doing and using a doctorate. Also, we encourage you to read other books that provide information about doing and using a doctorate. In particular, we recommend the books by Hayton, Gosling/Noordam and Phillips/Pugh (see Refs. [5, 6, 7], Further Reading).

We also provide a tool to help you in your decision-making process. This tool is available on the book's website (www.ADoctorateAndBeyond.com).

1.5 Further Reading

[1] A. Flexner "The usefulness of useless knowledge" Harpers, Issue 179, June/November 1939.
[2] D.K. Sokol "Is a PhD the right action for you?" Guardian Professional.
[3] The Thesis Whisperer "What to say, when someone asks you: Should I do a PhD?" The Thesis Whisperer, November 2011.
[4] M. Aliotta "Ten good reasons for doing a PhD" Academic Life, November 2011.
[5] J. Hayton "PhD - an uncommon guide to research, writing and PhD life" published by James Hayton, 2015.
[6] P. Gosling, B. Noordam "Mastering your PhD: Survival and success in the Doctoral Years and Beyond" 2nd edition, Springer, 2011.
[7] E.M. Phillips, D.S. Pugh "How to get a PhD: A handbook for students and their supervisors" Open University Press, McGaw Hill Education, 4th edition, 2005.

Chapter 2
How Might You Use Your Doctorate, if You Get One?

2.1 Overview

This chapter will give initial ideas on where you can use a doctorate if you obtain one. To put some structure into this topic, we will examine the university, private and government sectors separately. However, in truth, there is much overlap. It is also possible to move between these sectors. Specific examples of people who have successfully made the transition in both directions will be provided later in the book.

2.2 Positions that May Require a Doctorate

2.2.1 University Employment

> A doctorate is a necessary prerequisite to gain employment in a university.

If you seek employment in a university as a fresh doctoral graduate, then your options would include the following:

- a postdoctoral position (possibly abroad) involving research on an externally funded project
- a junior faculty position including teaching and/or research.

In the long run, your options within the university sector include either specializing or diversifying your research interests and ultimately becoming a full professor. You might move into administration or management positions such as Dean of a Faculty or even President of a university.

To illustrate, Graham, upon finishing his doctorate in Australia, took up a junior faculty position at Imperial College (part of the University of London) in the United

Kingdom. He later returned to Australia where he moved through the academic ranks to full professor. He changed his research focus repeatedly. Indeed, Graham aimed to "reinvent" his research direction every five years. This was challenging but also refreshing and, ultimately, rewarding. He was a regular faculty member for a quarter of a century, then became Dean of a Faculty and ultimately returned to a full-time research position.

2.2.2 Private Sector

If you are a fresh doctoral graduate wanting to work in the private sector, then, in the short term, it is probably best to apply for a position where your specific field of expertise is relevant. However, eventually you may be given the choice to either stay in the same technical stream or move to another technical area. Alternatively, you may choose to progress to a more managerial or leadership role.

To illustrate, Stefan began his industrial career by becoming head of the advanced control department in a refinery. This position was close to his field of technical expertise as he had previously worked as an Associate Professor in the area of advanced control engineering. Once in industry, he increasingly built his management career and became CEO for an aviation joint venture company. Expanding on the business and strategic skills he acquired in this position, he diversified to becoming the chief strategist for the international part of the corporation. He then served on the board of an aviation venture company and as CEO of a crude pipeline company. On leaving that company, he moved to South Africa where he started a company which offers personal and business coaching. The latter job combines all of his earlier acquired skill sets.

Many of the doctoral graduates we have known ended up working in industry. The jobs are many and varied. Some of the positions are

- Designing and running Induction Heating Furnaces.
- Managing a major egg producing company.
- Being a senior process engineer in the Sugar Industry.
- Working on Marine Engine design.
- Working as an Advanced Process Control engineer.
- Acting as a Process Control consultant.
- Working for a Mining Company.
- Working for a state-owned industrial R&D institution.
- Working (indirectly) for a major mobile telecommunications company.
- Working (directly) for a mobile telecommunication company.
- Running the Computer Operations for a company involved in gambling.
- Designing a new generation of electric vehicles.
- Working for an investment company.
- Designing supply chain systems.
- Writing software for train scheduling.

2.2 Positions that May Require a Doctorate

Not all of these jobs were directly related to the specific topic of the person's doctorate. However, the doctoral training gave the person the skill to tackle a difficult task and bring it to completion. The doctoral training was thus a crucial aspect of their career trajectory.

To further illustrate how a doctorate can be used in industry we quote the case of Meng Wang who has a position within Ericsson AB in Stockholm, Sweden. He commented as follows:

> "A doctoral degree can be of great importance in industry, in particular for those companies that emphasize and invest in research".

"For the part of Ericsson Research for which I work, more than 70% of the employees hold doctoral degrees from Sweden or overseas. This proportion is even higher for new recruits. Compared to other departments, the jobs within the research department are more conceptual or theoretical. The research methodology is similar to academia but is more focused and has more limitations. For example, how to further cancel some type of interference for a given specific telecommunication system, or how to predict the performance if one were to add some new feature to the standard. On the other hand, in other departments (such as system design or development), a doctoral degree may not be so important. These latter departments focus more on implementation issues, for example software development. In these situations, high level programming skills could be more important than a doctoral qualification".

2.2.3 Government Sector

Other opportunities for doctoral graduates lie in the government sector. The opportunities in government can include government research laboratories or government accreditation offices. Doctoral graduates are also highly sought after by procurement agencies to research and advise on new defense technologies. Another avenue for a doctoral graduate in government is as a legislative advisor.

As an illustration, we mention the career of Julio Braslavsky. After completing his doctorate, he first worked in a university-based research group focusing on R&D projects connected with the mining industry. He later joined a government research organization in Australia (CSIRO) where he manages a research team working on renewable energy.

2.2.4 A Private Consultant

A doctorate often provides high-level expertise in a particular applied area. This expertise can then be exploited by becoming a private consultant. We can quote many examples of people who pursued this career choice, e.g. one who became a consultant in environmental remediation and another who is a consultant in anodic protection.

Consultants offer high-level advice to industry. They typically lead customers to articulate their problems and then to find solutions. Training at the doctoral level can be invaluable in helping a customer understand their difficulties and to find solutions.

2.3 Skills and Passion

As we have seen from the previous section, there exist a wide variety of options where one can use a doctoral qualification. One of the reasons for this wide diversity is that a doctoral degree goes beyond your specific specialization and includes the acquisition of other skills that will help you as your career progresses. These skills include

- perseverance
- networking
- the ability to ask deep and meaningful questions
- the ability to analyze data
- the ability to see things in data that others have overlooked
- the ability to convince other people about the value of your ideas
- the ability to organize material so that others can readily understand and appreciate it

Doctoral training includes tackling a hard task and seeing it through to completion.

A good compass to find your personal way amongst this amazing variety of options is to follow your passion. This may lead to surprising twists and turns.

For example, Rowan Atkinson started his career by completing a Masters of Science Degree with the title "The application of self tuning control", in 1978 at Oxford University. However, he did not continue with a doctoral qualification. He moved from an engineering degree to the field of entertainment. Indeed, you probably know Rowan Atkinson as "Mr. Bean" or "Black Adder".

Conversely, Dr. Brian May, the brilliant musician and lead guitarist of the rock band Queen began a doctorate in the 1970s. He finally completed his doctorate in Astrophysics in 2007 at Imperial College, London. He was a collaborator on the New Horizon Pluto Mission with NASA and served as Chancellor of Liverpool John Moores University from 2008 to 2013. We will include a specific contribution form Brian later in the book—see Sect. 9.2).

As another example, some of our readers may have heard Professor Brian Cox (A Royal Society Research Fellow at the University of Manchester) present his wonderful science shows for the BBC. Interestingly, Cox was also the keyboard player for the pop group D:Ream. He completed his doctorate in particle physics after D:Ream disbanded.

> The trick is to find what you love doing because this is where you will both excel and find satisfaction.

This was beautifully articulated by Steve Jobs (former CEO of Apple) in his Stanford University commencement speech on June 12, 2005 (see Ref. [1], Further Reading).

We encourage our readers to listen to this speech or to, at least, read it.

2.4 Summary

In this chapter, we have described a number of activities and careers that doctoral graduates can choose. Do any of them inspire you? Could you see yourself passionately engaged in any or perhaps several of them?

Remember that many doctoral graduates move across several fields during their career (see Sect. 2.2.3). So, for now feel free to identify with, or feel inspired by, several of the options we mention

- University Employment (see Sect. 2.2.1).
- Private Sector (see Sect. 2.2.2).
- Government Sector (see Sect. 2.2.3).
- Private Consultant (see Sect. 2.2.4).

For perspective, about 45 % of Graham's doctoral students have chosen the university sector, 45 % the private sector and 10 % either the government or consulting options.

Recall that we offer a decision-making tool on the book's website. This is intended to further focus your decision-making process.

2.5 Further Reading

[1] S. Jobs "Stanford University Commencement Speech, June 2005" Stanford Report, June 2005.
[2] Australian National University "PhD students: Careers outside academia" ANU Careers Centre.
[3] P. Gosling, B. Noordam "Mastering your PhD, Chapter 21 Putting it all Together: a PhD, So What's Next?" Springer 2nd edition, 2011.

Chapter 3
The Where, Who and What of Doing a Doctorate

3.1 Overview

This chapter addresses three core questions that accompany your doctoral deliberations, namely the "where", "who" and "what", of the process. More specifically,

- Where should you do it?
- Who should be your supervisor?
- What should you work on?

Depending on your interests, you may decide to begin with any one of these questions to initiate the process. Ideally, your answers to these three questions need to be aligned. Since you can begin with any of the questions we will treat each of them as if the other two were as yet undecided.

3.2 Selecting an Institution

In determining the best place to pursue your doctorate we examine two related questions namely what attributes to look for and how to check those attributes.

3.2.1 What to Look for?

You need to think carefully about the kind of environment that will be best for your doctoral studies. In particular, you will not be surprised to hear that (Fig. 3.1)

> It is hard to do a doctorate in a vacuum.

> We believe that the optimal place to do your doctorate would contain many, if not all, of the following attributes:
>
> - A collection of top academics
> - High-powered postgraduate courses
> - Interactions with other departments in the University (and industry, if you are interested in applied work)
> - A stream of top-level academic and industrial visitors
> - A stream of postdocs from around the world
> - Other researchers working on topics of interest to you
> - Funding for doctoral students including stipend, equipment, conference travel, etc.

Regarding the question of whether to stay at the same institution where you did your undergraduate degree or to move elsewhere, we simply say that this decision should be informed by considering the above list of requirements: If you can change to an institution that scores higher on this list, then it is advisable to do so; if your undergraduate institution already scores high, then a change may not be necessary.

> Students who are seeking to ultimately work in industry should give particular consideration to doing their doctorate in a University Group that has strong links to industry. The industrial connections acquired during your studies will stand by you as your career evolves. Indeed, doing a doctorate on a topic closely aligned with industry can often be viewed by companies as a protracted job interview.

3.2.2 How to Check These Attributes

Using online tools it is relatively easy to check an institution's research "readiness". A first step is to find the names of the Professors using a search engine. Then google their names. Check their research interests, publications, coauthors, etc. Also, check if they regularly attend conferences and note the people with whom they collaborate.

Then look for the leading conferences in your area of interest and see if the particular professors that you are investigating attend and whether or not they give plenary or key-note talks. Then match the key people back to an institution.

It may also be helpful to contact current doctoral students and postdoctoral fellows at an institution. You will find that many of them will be happy to share their impressions and experiences with you.

Fig. 3.1 "It is hard to do a doctorate in a vacuum"- see Sect. 3.2.1

In summary,

> Check the research environment of your intended institution. Is it doctoral fit?

If you do not achieve acceptance at your institution of first choice, or personal reasons prevent you from going there, then move onto your second or third-best choice and possibly think of changing later. (See also the comments made in Sects. 7.4, 7.5 and 7.6).

3.3 Selecting a Supervisor

In this section, we assume that you may have already chosen the institution but do not, as yet, have a specific supervisor. We also assume that you have not, as yet, locked down the topic.

Things to look for when selecting your supervisor

- Is the potential supervisor active in an area that interests you?
- Does your potential supervisor have a strong publication record?
- Is your potential supervisor well-connected to the international academic community in your chosen area?
- Does your potential supervisor have strong financial support for their research?
- Is your potential supervisor up-to-date with recent developments in your chosen field?
- Do you feel you would have a good rapport with the potential supervisor?
- Does it seem that the potential supervisor would have the time to supervise you properly?

The best way to check your potential supervisor against the above list of desirable attributes is to visit the potential supervisor. It will also help if you talk to his/her current students and postdoctoral fellows.

The above considerations probably point you in the direction of a senior academic. Indeed, that is very often an excellent choice. However, there are caveats. For example,

> Make sure that your potential supervisor has time to spend with you!

The first author recalls visiting a top-level research institution where a famous researcher gathered his multiple doctoral students together. He said to the students, "I am sorry I cannot remember all your names so please introduce yourselves". Did this imply that the person was overcommitted? We leave you to think about this.

3.3 Selecting a Supervisor

We believe that

> Doing a doctorate is as much learning *how* to do research as actually doing it! So choose a good role model for your supervisor.

One option worth considering is to work with a younger academic who is full of enthusiasm and who is "on top of" his/her field. Of course, even better, when this young person works in a group which also contains senior people. It is highly likely that

> Senior researchers set the tone of the research environment.

Also be sure to check that your potential supervisor is willing and able to work freely and openly with others. Remember that your supervisor will be your close colleague and mentor for a long period so it is wise to choose one with whom you believe you can establish a good rapport.

We quote Dr. Roger Davies who was introduced earlier—see Sect. 1.2.4, p. 5, "I chose to leave Leeds University to undertake research at Swansea University. I chose the subject because Professor Richardson, my supervisor, was recognized as a global giant in the Chemical Engineering world. He was also able to wangle a Science Research Council grant for the work. I got lucky. He was also a great guy. Our working relationship was business-like with mutual respect".

Both authors of this book were fortunate to have very engaged supervisors. In Graham's case it was Brian Speedy who had an amazing ability to inspire people and to see practical links for more theoretical research. In the case of Stefan it was Torsten Bohlin who had a wonderful grasp of estimation theory and its application to real-world problems. Both Graham and Stefan ended up doing further work with their supervisors: Graham coauthored a book with Brian and Stefan's thesis work was developed into a commercial product.

On the other hand, both authors have seen other supervisors who treated their doctoral students in a less than optimal fashion. We have heard some supervisors imply that their students were stupid because they could not understand or remember something. Luckily, vast knowledge is never a prerequisite for doing a doctorate as Dr. Carl Sagan (American Astronomer, Writer and Scientist, 1934–1996) said (see Ref. [5], Further Reading)

> "Knowing a great deal is not the same as being smart; intelligence is not information alone but also judgement, the manner in which information is collected and used".

Another factor to think about is breadth versus depth. A very well-known Professor of Chemical Engineering that we know once said,

> "There are supervisors who know everything about nothing and those who know nothing about everything".

The interpretation of this slick sounding sentence is that there are people who are very narrow in their knowledge and others who are very broad. Which is best? Certainly, in order to be at the very top of one's field, one needs to be sufficiently "focused", i.e. an element of narrowness is inevitable. Also, while doing your own doctorate you will need to have complete mastery of the area and that inevitably means some degree of narrowness. On the other hand,

> Big breakthroughs often come by combining ideas from different areas.

Hence it is always good to keep a broad perspective. So, the question of breadth versus depth is not a right or wrong issue but one of striking the right balance. When choosing a supervisor be wary of people who are very narrowly focused, since this may mean that they are inflexible and locked into a particular way of thinking.

The right balance might be a narrowly focused supervisor in a broadly focused Department or group.

Our key message is (Fig. 3.2)

> Choose your supervisor wisely. Remember that if they are not heavily involved in research themselves, then they will be hard pressed to inspire or guide you.

3.4 Selecting a Topic

Typically there are two scenarios with regard to topic depending upon whether you have a predisposition towards a specific area or you do not. The former may occur, for example, if you are sponsored by industry. In this case, the key thing is to find the best supervisor and location to match the chosen topic.

If you are open to the selection of a topic, then the choice is probably best done in conjunction with the choice of supervisor and/or location.

A particular question students often ask is "Should I choose a topic that is currently "hot" in the field, so I can publish more easily, or a topic that might be less "hot" but which might align better with my own interests or those of my preferred supervisor?"

Fig. 3.2 "Choose your supervisor wisely. Remember that if they are not heavily involved in research themselves, then they will be hard pressed to inspire or guide you". - see Sect. 3.3

We believe that the best choice for your topic is one that combines all three elements, i.e. is one which
- Is of great personal interest to you.
- Is well aligned with your proposed supervisor.
- Is currently a "hot" topic.

Indeed, we would go so far as to suggest that the absence of any one of these components will weaken the benefit of your doing the doctorate. However, inevitably compromises need to be made. In such a case, we believe that the correct priority is to weight the elements in the order listed above, i.e.

The most important aspect of choosing a research topic is that it inspires you.

We give least weighting to the topic being "hot". Our reasoning behind this recommendation is that, if you get a breakthrough in a "hot topic", then you will gain high exposure. However, "hot topics" inevitably attract a lot of competition including work by established researchers. Thus the "all or nothing" aspect may make this a risky undertaking.

We can quote the case of Arie Feuer. Arie spent most of his career at the Technion in Israel but he also founded and became Chief Technical Officer for a spin-off company. His doctorate was conducted at Yale University. His supervisor set him a topic which amounted to resolving a problem that had remained open for two decades. Fortunately, Arie and his supervisor succeeded in this challenge but the associated risks are clear.

Quoting Thomas Schön (see Ref. [4], Further Reading)

"Research is by definition a high risk and uncertain activity.

If it were possible to perfectly plan and exactly predict the outcome this would mean the results were known and hence it would not be research".

We have many examples of the power of getting the right question.

Graham was once contacted by a student. He was working with another supervisor at the time. The student said "I have been doing my doctorate for three years but I do not think I am making good progress". Graham asked the student to come back on Monday and give him a three-minute synopses of his work-to-date (see Sect. 11.3). The student came back on Monday and said "I know I only have three minutes but, in all honesty, I have nothing to tell you!" After considerable discussion Graham and the student agreed that the source of the difficulty was that the student did not have a good problem. So they worked together to choose a better problem! Exactly one year later, the student submitted his doctorate which received excellent reviews.

Selecting a good topic is not as easy as it sounds especially if all your experience has come from your undergraduate education.

As one of our doctoral students said

> "A good undergraduate student gives great answers to questions, a good researcher asks great questions".

However, you need to try to get it right, since

> Having a good problem is crucial if you want to end up with a good thesis.

3.5 Summary

In this chapter, we have addressed the three important questions, relating to

- Selecting an Institution where you can do your degree (see Sect. 3.2).
- Selecting a Supervisor (see Sect. 3.3)
- Selecting a Topic (see Sect. 3.4)

Do not worry if you do not have a definitive answer to any of these questions right now. It is early days in your decision-making and the issues are a "stock-take" of where your thinking has reached.

Our general advice for selecting institution, supervisor and topic is think deeply about the issues discussed in bullet points listed in Sects. 3.2.1, 3.3 and 3.4.

These are generic lists that you will necessarily have to adapt to your personal interests. For example, you may be particularly interested in applied work, theoretical work or a certain field such as automotive or medical applications. Clearly this will influence your choices.

3.6 Further Reading

[1] A. Greenspoon, "9 Things you should consider before embarking on a PhD" Elsevier, April 2013.
[2] J. Peironcely, "10 Things you should know before starting a PhD" Next Scientist, Graduate School Advice Series.
[3] T. Brabazon, "10 Truths a PhD supervisor will never tell you" Times Higher Education, July 2013.
[4] T. Schön, "How I supervise PhD students" Pedagogical Seminar, Uppsala, Sweden, April 2015.
[5] C. Sagan, "Cosmos" Little Brown Book Group, UK, 1980.

Chapter 4
How Hard Will You Have to Work?

4.1 Overview

This chapter addresses the issue of the work load associated with doing a doctorate. Some students approach the task of doing a doctorate as a 35 hour a week job! We have some bad news especially for doctoral students in Engineering and the Physical Sciences. The truth is that it is a 50 to 60 hour a week job. On hearing this, many potential students (and, indeed, sometimes also their supervisors) express concern—this is not realistic. However, we need to explain a little further, so please read on.

4.2 An All Consuming Activity

Even if you are not sitting at your desk for 10 hour a day, your mind should always be turning ideas over. (Fig. 4.1)

> Research can, and should be, all consuming.

Indeed, it is often in, so-called, idle moments, that the critical idea suddenly arises. The only risk is that, at times, research can be so exciting that one forgets how much time one has devoted to a particular problem. This is discussed in the interesting article by Abraham Flexner (see Ref. [2], Further Reading).

4.3 Things to Avoid

Actually, much has been written on doing a doctorate. The following book contains many useful ideas and suggestions: "How to Get a Ph.D.: A handbook for students and their supervisors", Estelle M. Phillips and Derek S. Pugh, Open University Press, McGraw Hill Education 4th Edition 2005 (see Ref. [3], Further Reading).

Fig. 4.1 "Research can, and should be, all consuming" - see Sect. 4.2

We particularly recommend the section on how not to get a doctorate which emphasizes the intersection between your passion and vision and their intersection with the importance of the guidance you get from your supervisor.

4.4 Progress

Doing a doctorate can be extremely rewarding. However, it can also be extremely frustrating. A typical representation of the progress of a doctorate versus time is shown in Fig. 4.2.

Doing a doctorate can also be an incredibly messy business and students can easily become despondent. Dark periods are inevitable. Graham recalls one of his doctoral students who had a very poor self-image. After several years trying to do doctoral work, he came to Graham and said, "Sorry, Graham, I'm not up to this job!" After a lot of discussion, Graham finally said, "Okay, I agree let's suspend the doctoral

Fig. 4.2 "The worst times can present the best opportunities, so don't give up" - see Sect. 4.4

work". "Instead, would you like to help me sort out some ideas that I need for a masters level class I am teaching?" The student agreed and began compiling some notes and doing some simple examples. During the course of this work, the student came back and said, "I feel even more stupid than before! The simple examples that you asked me to do will not work the way the theory predicts". Graham quickly replied "Don't panic, let's see where you are going wrong!"

Guess what? It was the published "theory" that was wrong (or, to be more precise, it was incomplete). The student was able to write two brilliant papers correcting the established wisdom in the literature. These papers were accepted by one of the top journals in our area. Remember

> The worst times can present the best opportunities, so don't give up!

4.5 Impact on Your Family

Regarding how hard one has to work and the impact on other family members, it is relevant to quote the story of Ramon Delgado who was one of Graham's doctoral students from Chile. After deciding to do a doctorate he found that his wife was pregnant. Not only pregnant, but with twins! He decided to still come to Australia to do the degree but there were significant issues surrounding moving to a foreign country, financial constraints, support for family and available time. Initially Ramon gave priority to his family but this meant that the degree progress was negatively impacted. Towards the end of his programme, he took a reality check and decided to focus more heavily on the doctoral work. He was then able to complete the programme. However, the journey was a difficult one for Ramon and his family. He offered the following advice to others.

"If your are thinking of doing a doctorate and you have a young family that is dependent upon you, then I strongly recommend that you think very carefully. You will definitely face difficulties. If you decide that the ambition to get a doctorate is extremely strong and you are willing to pay the price in terms of disruption to your family then choose your supervisor carefully. Think about whether or not your supervisor has previously had successful outcomes with people in your situation. Also try to gauge whether or not your supervisor has access to adequate research funds that can support your extra financial requirements and is willing and able to allocate them to you. You also need to have a clear view regarding the balance you are prepared to make between work and the needs of your family".

4.6 Summary

In this chapter, we have highlighted the intensity of commitment that the pursuit of a doctoral degree in engineering and physical sciences requires. For the purpose of summary we particularly emphasize the words of wisdom or how *not* to get a doctorate (see Sect. 4.3).

All the commitment in the world will likely not prevent you from experiencing times of stagnation, frustration and despair. For these times, we offer words of encouragement, provided these trying times are not just due to lack of commitment and provided you have built good rapport with your supervisor and peers as suggested in earlier chapters, then you will get through!

If you are using the decision aid tool, then we refer you again to the book's website.

4.7 Further Reading

[1] T.B. Schön, D. Broman "A Process Framework for Agile Doctoral Supervision", in preparation, communicated to authors.
[2] A. Flexner "The usefulness of useless knowledge", Harpers, Issue 179, June/November 1939.
[3] E. Phillips, D.S. Pugh "How to get a PhD: A handbook for students and their supervisors" Open University Press, McGraw Hill, 4th edition, 2005.
[4] P. Gosling, B. Noordam "Mastering your PhD" Springer, 2nd edition, 2011. Chapter 7 "Dealing with setbacks".

Chapter 5
How Long Will It Take You to Get a Doctorate?

5.1 Overview

This chapter addresses the question of the length of time that it will take to finish a doctorate and whether or not it can, or should, be done part-time. Getting a doctorate requires enormous focus and hard work. Also, your supervisor may demand certain concrete outcomes, e.g. hardware to be built, new theory to be developed, papers to be written, etc. So, make sure that you are aware that this will not come without effort and dedication.

At this point, we also refer you to Sect. 23.4 ("Working Smarter not Harder") and Sect. 23.5 on how to interpret "time management" as "energy management".

5.2 Mission Creep

Depending on the topic and the person there tends to be a time frame in which extra time leads to extra quality. However, beyond that, extra time can simply result "in mission creep".[1] The natural time frame is 3 to 4 years with exceptional cases being as short as 2 years and as long as 5 years.

> The key is to "roll up one's sleeves" and get stuck into the work.

Of course, the progress is never linear. Indeed, the reader is reminded of the "Progress versus Time" curve shown in Chap. 4. However, with focus and determination, it always seems to work out.

Graham was fortunate in his career to supervise some truly gifted and focused individuals. Some, but not all, of these are mentioned in this book. One that deserves

---
[1] Mission creep: Endless refinement of a project or endlessly widening of scope.

mention at this point is Rick Middleton. He was a brilliant undergraduate student. He was able to complete all of the work necessary to obtain a doctorate in just 2 years of study. So, it can be done!

5.3 Be Focused

Graham recalls one of his other doctoral students (Mario Salgado) who is mentioned at several points in this book. Mario was a mature student who already held a position as an academic in his home country of Chile. So, Mario uprooted his family and traveled half way around the world to do a doctorate in Australia. On his very first meeting with his supervisor, Graham, Mario said, "I am here to do a doctorate, but, I must finish the work and submit my thesis in exactly three years from today!"

Did this mean Mario cut corners in his degree? Absolutely not! Indeed, Mario produced a truly outstanding thesis and easily met Graham's (and, perhaps more importantly, his own) expectations. Moreover, he left Australia to return to Chile exactly three years after he arrived. Of course, he worked very hard and was able to focus his energy.

> If you are doing a doctorate it is a good idea to have a clear goal and finishing time in mind.

The above may actually sound quite straight-forward. If one of our young readers does a quick online search on "goal-setting and achievement" they may be surprised to see how vast the associated literature, bloggs and advice are. The reason is that goal-setting and deadline management are quite difficult skills to master in practice.

One currently popular approach bears the acronym of S.M.A.R.T. goals. An early mention of S.M.A.R.T. goals is the seminal contribution by G.T. Doran (see Ref. [2], Further Reading). Since then, there has been a continuous stream of publications sharing experiences and exploring variations; we encourage the interested reader to explore the literature.

> S.M.A.R.T. stands for goals that are
>
> **S**pecific
> **M**easurable
> **A**chievable
> **R**esults-focussed
> **T**ime-bounded

Some authors use slightly different words for the acronym with similar meaning. An extension of the S.M.A.R.T. idea creates S.M.A.R.T.E.R. goals by adding

> **E**thical
> **R**ecorded

We will discuss the S.M.A.R.T. system further in Sect. 16.3. This system is particularly useful when it comes to goals involving small to medium amounts of uncertainty. We encourage doctoral students who are reading this chapter for the first time to jump ahead and skim the section devoted to S.M.A.R.T. goals.

5.4 Goal Setting

All research, including doctoral research, involves high levels of uncertainty. So it is very important that you set yourself concrete goals. For example, you might set yourself the goal of having carefully read a particular research publication by next Sunday. With enough hours you can be reasonably certain that this goal will be achieved.

Understanding a research publication may take longer. Indeed, multiple hours of effort and extreme will-power unfortunately do not always guarantee that you will understand a particular scientific result. Similarly, proving mathematical theorems or solving a practical problem can be very difficult. Time investment alone does not guarantee success in achieving the desired goal by a desired deadline. Thus, you need to subdivide the task into intermediate goals that you know are achievable.

For example, the simple goal: "I want to have proven the mathematical theorem over the next four weeks", might sound like a good goal and allow you to fall asleep that night: after all, you have set a goal which, if attained, will be a significant milestone.

However, in this simple formulation, it is an "all-or-nothing" proposal. It is based on an investment of time. Alas, no matter how diligently you spend your time on the problem, it is not guaranteed that you will find a solution.

So a more subtle goal formulation would be to say: "Prove theorem within next four weeks; if no substantial progress after one week, get advice from fellow students: if still no substantial progress after two weeks, get advice from supervisor", etc. You could embellish this with: "Make an appointment with my supervisor after ten days, to discuss progress and to obtain advice".

Goal-setting is an art. We strongly advise you spend time refining this art.

> The key is to become more subtle about goal setting and to incorporate a spectrum of outcomes.

5.5 Can, and Should, You Do a Doctorate Part-Time?

A question that some students ask is can, and should, a doctorate be done as a part-time, say, whilst working in industry. There is no doubt that this is possible especially if the workplace is very supportive of the program. Indeed, some industries who are heavily involved in research and development might view your doing a doctoral degree as an integral part of your work function.

Other students simply do not have a choice other than to tackle the challenge of working in an unrelated field whilst completing a doctorate at the same time. In many countries this is more the norm rather than the exception. So if you are forced to face this challenge, rest assured many have previously successfully completed this journey.

Professor Richard Middleton (mentioned earlier in this chapter) said he had recently supervised a very successful part-time doctoral student in industry. However, he then noted that the industry strongly supported the student doing the degree and also the student's immediate supervisor in industry already held a doctorate himself.

> In general, we advise that, if your company wants you to do a doctorate part-time, that you make sure that the company understands and supports you in this endeavor.

Otherwise you should consider seeking leave of absence and finishing the thesis as quickly as possible. Then you can go back to your company and apply your new skills.

5.5.1 Balancing Work and Thesis

We have known many part-time doctoral students and, in every case, the students struggled to maintain momentum in the combined responsibilities associated with their job and research. At the end of the day, both suffered. This is especially so if the link between doctoral topic and work responsibilities is weak. So,

> If you want to do a doctoral degree, then, if at all possible, do it full-time. It is difficult to manage the dual demands of your boss at work and your degree supervisor.

As we were writing this section, a student (Raheleh Nazari) came in the door having just submitted her doctoral thesis. Raheleh did the first two years of her studies full time and then transferred to part time for financial reasons. She made the following comments, "Concurrently satisfying the demands of both my industrial position and those of my doctoral studies was the biggest challenge of my life! The industrial position required a different mind-set, even a different form of dialogue. The industrial position required rapid response to evolving events whilst the doctoral studies required a long-term vision. The necessity of switching between these two roles was very demanding. I would caution others from taking this route unless it is necessary. However, I want to stress that I would do it all again in a heart beat. It was certainly challenging but also hugely rewarding. It completely changed my life".

We have asked many supervisors of students doing a doctorate part-time and they all say that it is extremely difficult save when the company allows the student to view a significant part of their job as thesis related.

As Fred Brooks said (see Ref. [3], Further Reading)

> "Question: How does a large software project get to be one year late?
> Answer: One day at a time!"

5.5.2 Making a Success of a Part-Time Doctorate

However, some students choose to do a part-time doctoral degree because they are encouraged by their employer to do so or already have significant financial commitments, e.g. family, mortgage on a house, etc. These typically demand a higher level of financial support than is normally available to a doctoral student. If this relates to your circumstances then the following reflections from Dr. Steve Mitchell, who did his doctorate as a mature, part-time, student may be helpful.

"In my case it was imperative that I maintain a steady professional income for the full duration of my doctoral degree in order to meet my growing financial obligations. The reality of this caveat was that study inevitably required a part-time approach, and part-time study brings additional challenges to bear; was I really prepared to give up a significant proportion of my weekends for the next eight or so years?

Another challenge with the part-time approach is the intermittence of suitable opportunities to commit to your research program. It can, at times, be very difficult, if nigh impossible, to consistently complete a full working day, return home to embrace your domestic responsibilities, and then, when the rest of the house slumbers, to recommence your research. This inevitably pushed most of my research opportunities into the weekend, but it can be a challenge to sustain your problem solving momentum when several days may have elapsed between research efforts.

Sustaining an appropriate work-life balance under the described circumstances can be complicated. Working consistently across weekends requires a reallocation of priorities. My priorities were family first, research second, and household chores a distant third. This meant letting a lot of the non-essential household maintenance activities fall by the wayside for literally years at a time; this is stated without even a hint of exaggeration. As a result, a vital component of any successful attempt at a "Mature Age Doctorate" is a very understanding and supportive partner.

At the completion of my doctoral degree I felt a tremendous sense of accomplishment. I had achieved something of real personal significance, something that I now hold very close to my heart; a twinkle in my eye. It is also a significant achievement for my family. For my parents, it is a celebrated accomplishment that cannot be overstated. For my children, it is an unspoken declaration of the value that our family places on education and lifelong learning, irrespective of the career direction they choose to take into the future.

A doctorate can be a key which may open the door to an academic career. A postdoctoral research position at another institution would be the next logical step along the academic path. A few dedicated years of concentrated effort to expand and further develop your research, build up a solid publication record, seize the opportunity to expand and nurture a valuable network of collaborators in your field from different parts of the world. However as a Mature Age academic you are arriving rather late to the event. In my case I now had four children, three of whom were settled in local schools, and a spouse with a developing career of her own. Packing up my family to take up a postdoctoral fellowship elsewhere was not a viable proposition in either the geographical or financial sense.

I was very fortunate to acquire an industry sponsored contract as an academic at my local university. This provided me with the wonderful opportunity to both lecture and conduct research for several years. However, with a still maturing publication record in the highly competitive international marketplace of academics seeking tenured positions, securing a tenured position has been difficult. This has left my academic career in a rather volatile position, dependent upon the relatively capricious nature of short-term industry funding. It currently appears inevitable that funding gaps will, at some point, force my hand to leave academia. The decision that needs to be made is whether to leave on my own terms as soon as a suitable role outside of academia appears, or whether to ride the funding wave and see if an unforseen opportunity arises. For the sake of stability for my family, the former is the most likely scenario.

Reflecting on my decision to undertake a Mature Age Doctorate, I am of two minds. One part of me feels incredibly proud of my achievement and, neglecting the volatility of my position, I have thoroughly enjoyed my experience as an academic. It has enabled me to grow and to be challenged as a professional, to impart my knowledge and passion for learning onto my students, and to present my research via conferences at amazing locations around the globe."

As an update to the above contribution, we note that Steve recently left the academic world to join the electronics division of a large company. In that role, he uses the skills he learned as a part-time doctoral student to organize work for others and to seek innovative and elegant solutions to tough design problems. So, Steve's doctoral experience turned out to be positive in a fashion that he had not previously imagined. Interestingly, he is also now actively encouraging students to pursue doctoral studies on topics of relevance to his company.

5.6 Summary

Whereas Chap. 4 focused on the quality of work and your commitment, this chapter has examined the issues of quantity of work, i.e. how much time you will need to commit to your doctoral work. Even a lot of time spent on your doctorate can become too little if you succumb to mission creep, lack of focus or procrastination, including writer's block (see Sect. 5.2 on mission creep and Sect. 5.3 on focus). Completing a doctorate in engineering or physical science on a full-time basis will usually take 3 to 4 years, with 2 or 5 years somewhat more exceptional.

However, many students do not have the luxury of being able to study full-time. This could be due to financial constraints: you may have to work part-time to earn a living or support family. If you find yourself in this group, then Sect. 5.5 gives words of advice and encouragement.

5.7 Further Reading

[1] P. Gosling and B. Noordam, "Mastering your PhD" Springer, 2nd edition, 2011.
[2] G.T. Doran, "There's a S.M.A.R.T. way to write management's goals and objectives", Management Review AMA Forum **70**(11):35–36.
[3] F. Brooks, "The Mythical Man-Mouth", Addison-Wesley, 1975.

Summary of Part I

Part I of this book has been concerned with the decision process centered on whether or not a doctoral degree is the right course of action for you. This question has been addressed at two levels: In a first examination of the question (Chap. 1) we took a high level view, asking whether you fall into any of the following four categories:

- you aspire to undertake a career that requires a doctoral degree,
- you are fascinated by innovation in a particular field,
- you are fascinated by the quest for science and engineering knowledge as an end in their own right, or
- you want to achieve a personal goal.

If any of the above applies to you, then the likelihood is that a doctorate is the perfect course of action for you.

To firm up the decision process, Chaps. 2–5 examined the questions at a deeper level and addressed the following topics:

- Examples of where you can use a doctorate, including universities, industry and government (Chap. 2).
- The importance of choosing a place, supervisor and topic and the alignment of all three (Chap. 3).
- An indication of how hard you will have to work (Chap. 4).
- A discussion of how long a doctoral degree might take and of the difficulties associated with doing a doctorate part-time (Chap. 5).

When combined with the initial indication, this more detailed information should help you arrive at a solidly founded decision. You can also use the decision aid tool on the book's web-site: A doctorate and beyond.

Part II
Doing Your Doctorate

Overview of Part II

So you have decided to do a doctorate. Wonderful news! Based on this important milestone in your life, we next discuss the process of actually doing and completing a doctoral degree. In particular, we address questions such as: what outcomes will be expected, student-supervisor interactions, the value of networking, tools of the trade, the art of publications, the art of making great presentations, the ethics of being a doctoral student and how to write a thesis. Clearly these are all important issues that you will need to address during your doctoral journey.

Chapter 6
How to Begin

6.1 Overview

This chapter is intended to get you started on the road to obtaining a doctorate. We address such issues as hitting the road running, maintaining records, keeping the final thesis in mind as you proceed and working in a group.

6.2 A Guiding Principle

You should keep in mind from the beginning that

> A doctoral programme should contain original ideas - a new contribution to knowledge or know-how.

So, how will success be measured? Typically, original ideas can be published and hence one simple criterion is to make sure that publications flow from the work. This can include a mix of refereed conference papers and refereed journal articles.

6.3 Hit the Road Running

We strongly advise that you should get into the real thesis work as early as possible. Be aware that the doctoral journey will often entail changes in the direction and re-evaluation of goals. Hence, the sooner you get moving the better placed you will be to cope with the hiccups along the way.

One of Graham's strategies, which he used with essentially all of his doctoral students, was to suggest a "warm-up" project. These projects were intended to establish an initial direction and to set initial student–supervisor rapport. The concrete goal of these warm-up projects was often to produce a first publication, e.g. a conference paper. Interestingly, about 50% of Graham's doctoral students essentially followed the initial direction whereas another 50% adapted and changed the original direction as the project evolved.

The idea of a "warm-up" project is very appealing to the author's of this book. If your supervisor does not suggest such a project, then you might consider asking him/her if they would be prepared to suggest one.

> Indeed, if your supervisor is not in the habit of suggesting warm-up projects but the idea appeals to you, suggesting one yourself, or asking for one, shows initiative and is an early opportunity for building rapport.

6.4 Keeping Records

A very important discipline that you should embrace from the very beginning is that of accurate record keeping. These records will prove extremely valuable as the work proceeds. This advice applies to all projects including those with an experimental or theoretical bias.

We suggest that you buy a note book that you can add extra volumes to over time. Alternatively, if you prefer, you can keep the records in a suitable computer-orientated format. Either way,

> The time invested in maintaining a "log-book" will pay huge dividends later.

Beyond this advice, you should be aware that there are formal guidelines for keeping records that most institutions insist upon. You should thus read Chap. 12 on the Ethics of Research as soon as possible.

6.5 Working in a Group

Some projects will require that your project be part of a larger team effort. This arrangement has many advantages but also contains some potential disadvantages. Some of the advantages and disadvantages are listed below.

> Advantages of having your project part of a larger team effort include:
> - Benefiting from mutual support
> - The excitement of a large project

- Possible financial support opportunities, especially if your project is linked to industry
- Shared goals
- Shared equipment and resources

Disadvantages of having your project part of a larger team effort include:

- Making sure you receive due credit for your individual contributions.
- Being sure you do not spend all your time writing progress reports for the funding agency.
- Ensuring that, at the end of the day, you will have a clearly identifiable separate contribution that you can "call your own" when it comes to writing your thesis.

We know of several real cases that encountered the kind of difficulties mentioned above. In one case, an engineering student decided to abandon his project because he could not find enough time in his busy work schedule to do his own research work. Fortunately, he was persuaded to have a straightforward discussion with his supervisor and went on to complete the doctorate. In another case, a medical doctoral student was unable to find anytime to do doctoral work due to pressures arising from a large grant held by the supervisor. Again a frank discussion with the supervisor resolved the problem.

There is an overlap between the issues addressed above and issues related to the ethics of research. So again, we advise that you read Chap. 12 on the ethics of research as soon as possible.

6.6 Working with Industry

A topic related to working in a group is that of working with industry.

Carrying out research in collaboration with industry can have many advantages including:

- Providing a project of contemporary and, possibly, commercial relevance.
- Providing funding.
- Providing motivation.
- Providing exciting feedback and relevance.
- Providing opportunities for future employment.

> However, there are also potential negative aspects that you need to be aware of including:
>
> - Being sure that you give some weight to your own individual contribution which may not overlap 100% with the industrial goals.
> - Being sure that you do not spend all your energy satisfying the industrial sponsor.
> - Being aware that industry may seek a short-term deliverable whereas your thesis may depend upon achieving a longer term research objective.
> - Making sure that the project satisfies the requirements of a doctoral thesis.
> - Issues with respect to intellectual property and publication rights.

It is always helpful to have a frank and realistic discussion with the industrial sponsor at an early stage of the work to establish the working arrangement.

In particular, the final bullet point regarding intellectual property and publication rights is a key issue to be discussed with an industrial sponsor. It is important that there be a mutually agreed view on this prior to beginning work so as to avoid the potential for future conflict and misunderstanding.

Questions that should be asked in relation to intellectual property include the following:

- Will there be an opportunity for patents?
- If so, what royalty issues need to be resolved?
- Will the industry place a ban on publication of the results and, if so, what is the time over which such restriction will hold.
- How, and where, can the results be published?
- Will industry require that parts of the final thesis be kept confidential? If so, does the University have appropriate mechanisms to ensure this can happen without jeopardizing the evaluation of the thesis?
- Will attendance at conferences be encouraged and how will the content of conference papers and presentations be vetted?

If you are contemplating doing a doctorate in collaboration with industry then you are strongly advised to discuss the above issues at a very early stage. The authors of the book speak with authority on this topic based on many positive (and a few negative) experiences.

6.7 Summary

Part II of the book, "Doing Your Doctorate", opens with fundamental considerations relating to getting started.

6.7 Summary

It contains three key messages:

- To be aware that your doctoral project needs to contain original ideas. You know that this is the case when publications (such as conferences or journal papers) flow from your work; see Sect. 6.2.
- Get started on your thesis work as soon as possible. A typical doctorate project will take three to four years and contain other obligations in addition to thesis work such as course work, part-time teaching and other distractions. It can be very tempting to postpone the start of you thesis work. We strongly recommend that you start early; see Sect. 6.3.
- Keep records (Sect. 6.4): What do you need to read? What approaches have worked and what did not? What are your conclusions? What ideas have lead to follow up work? Have you accurately recorded the results?

Finally, Sects. 6.5 and 6.6 contain points that you should keep in mind if you are engaged in a group project or a project conducted in collaboration with industry.

Chapter 7
Student–Supervisor Interactions

7.1 Overview

This chapter gives important advice regarding student–supervisor interactions. You are strongly advised to discuss this issue with your supervisor at an early stage so that an appropriate working relationship is established from the very beginning.

7.2 Engagement

Engagement with your supervisor is one of the most important aspects of doing a successful doctorate degree. If you loose contact with your supervisor then you cannot expect detailed help if you suddenly find yourself needing it.

In order to engage your supervisor you must first have the elements described in Chap. 3 in place. In an early meeting you should discuss the basis of the interaction with her/him. This includes the frequency of meetings. Also, after each meeting you should agree on the next meeting since the frequency can vary during the different phases of the work. As a rough guideline, a one hour meeting once per week is a good starting point but this could become much more frequent during periods of intense activity. At other times, one per month may suffice but we feel that this represents an absolute minimum frequency.

Good supervisor–student interaction is something like the interaction between a coach and a tennis player.

> The supervisor's job is to lob a well placed ball (i.e., an idea) over the net.
> The student's job is to hit it back harder than it arrived (i.e., to amplify the idea).

This works beautifully with some students leading to exponential "growth" in the ideas. With other students, the "ball is returned" poorly and exponential decay is the inevitable consequence.

> Students need to choose if they want exponential growth or decay to occur in the interplay with their supervisor.

The best students are able to keep their supervisors up-to-date with the project so as to gain maximal benefit from their supervisor's breadth of experience.

Students' time with their supervisor is precious so make best use of it. Be well prepared. Also, you should have exhausted other reasonable avenues for advice e.g. peers, literature, etc., before bothering your supervisor. Do not dump on your supervisor questions which he/she may believe were more reasonably your responsibility.

The following characterizations of different doctoral students follows the above suggested pattern of interaction:

> When a difficulty is encountered:
> - An average student goes to the supervisor and says: "I need help".
> - A good student goes to the supervisor and says: "I have encountered a problem but I see ten ways around it. What do you suggest?"
> - A very good student goes to the supervisor and says: "I have encountered a problem and I see ten ways around it. However, I think option 4 is best and I will look into it".
> - An excellent student goes to the supervisor and says: "I have encountered a problem and see ten ways around it. I think that option 4 is best and will look into it. Moreover, I feel this solution has much broader implications. Thus, I am keen to revisit other problems I have encountered".

In case the reader feels that the above characteristics are hypothetical, all supervisors meet these four classes of students. In this context, you may wish to quickly read Sect. 16.5 on the difference between an administrator, project manager, leader and entrepreneur. You will find remarkable parallels in the characterizations.

Finally, students need to be aware of cultural differences. This can be a severe impediment to free flowing communication between supervisor and students. There may be differences on whether or not students are prepared to challenge their supervisor. For example, Meng Wang (one of Graham's doctoral students from China, now working at Ericsson AB in Sweden) commented, "There can be a cultural difference/barrier between supervisor and students. People from different cultural backgrounds often deal with research and communication in different ways.

For example, doctoral students from some cultures can be more 'obedient' to supervisors and be less likely to challenge. Overcoming these cultural differ-

7.2 Engagement

ences/barriers in research is very important, since the efficiency of student–supervisor interaction depends upon good interaction and mutual understanding".

> Most importantly, be passionate about what you do and your supervisor will respond in kind.

7.3 Can I Work with Others as Well as My Supervisor?

There are several scenarios under which it can be highly advantageous to work with others. However, this needs to be done in an appropriate fashion. The key issue is to discuss the mode of interaction with your principal supervisor. We also suggest that there should be a good reason for working with others. Some reasons could be:

- The "bus syndrome" — there is always a chance your principal supervisor will be "run over by the proverbial bus", i.e. become unavailable for some reason. This is why many universities insist on there being a co-supervisor who is able to step up to the role of prime supervisor in the event that the principal supervisor takes a job elsewhere or becomes unavailable for reasons of health etc.
- If you are working on an interdisciplinary research topic. In this case it could be highly desirable to have multiple supervisors who have expertise in the different aspects of your project.
- There is a short-term visitor to your group who can add an extra dimension to your research.
- Your supervisor may make it a condition of taking you on that you work with others. This could arise, for example, if the prime supervisor has a heavy work load.
- You have a young supervisor but he/she suggests that you also interact with a more senior person so as to benefit from their experience and/or breadth of knowledge.

As an example of an interdisciplinary thesis we illustrate with Stefan's M.Sc. degree. As mentioned in Sect. 1.3, many universities allow (or even require) a Master's degree as an intermediate milestone towards a doctorate.

Stefan took this option and completed a M.Sc. on developing control algorithms for an implantable insulin pump for diabetics. This work was interdisciplinary between the University of California, San Diego departments of Systems Science (Control Engineering) and Bio-Engineering. Correspondingly, Stefan had two supervisors, one from Control Engineering (principal supervisor) and one from Bio-Engineering. On reflection, Stefan finds the following three points important:

- Interdisciplinary projects can be extremely rewarding, with all the satisfaction and challenges also set out in the Multidisciplinary Research Sect. 15.3.8.

- As a graduate student involved in interdisciplinary thesis work you need to ensure that you receive interdisciplinary supervision. All the previously mentioned issues of supervision still apply, yet now on a threefold level: the supervisor of the one discipline, the supervisor of the other discipline, and — indeed — the interaction of the supervisors creating the interdisciplinary guidance.
- It is important that the two supervisors are both enthusiastic and supportive of the interdisciplinary project. Stefan found, however, that the support of the second supervisors is crucial for an additional reason. Specifically, each supervisor tends to have better networks in their main field of study (control engineering and bio-engineering in Stefan's case). Thus, there are flow-on benefits to the establishment of networks from having dual supervisors.

7.4 Can and Should I Spend Time at Another Institution?

The issue of spending time at another institution is closely related to the issue of working with others.

We strongly recommend spending time at another institution during your doctoral studies if this can be arranged. Specifically, if your supervisor meets the basic requirements outlined in Sect. 3.2.2, then he/she should be able to make the appropriate arrangements. Indeed, it is highly likely that your supervisor will be in favour of you spending a period working at another institution during your studies.

> The advantages of spending a period elsewhere are manyfold and include:
> - Obtaining further opportunities for networking
> - Getting fresh input into your project
> - Seeing how other places operate
> - Obtaining access to particular equipment not available at your institution
> - Opening opportunities for future employment
> - Meeting other students from different cultural backgrounds
> - Living and travelling in a different country

Of course there are associated costs e.g. travel, arranging temporary accommodation in a new place, possible language difficulties etc. However, the benefits hugely out way the costs in almost all circumstances. Indeed, the advantages of spending time elsewhere can be so great that some institutions encourage combined doctoral degrees from two institutions. The latter option, of course, typically requires you spend approximately half of the program elsewhere. Your supervisor will be able to advise whether a combined doctoral degree is a desirable option for you or whether a better scenario is to spend a shorter period, say 6 months, at another institution.

7.5 Swapping Topic

Certainly adapting a topic almost always happens since the problem, as originally posed, may turn out to be intractable or too simplistic. Topic adaptation is usually achieved in collaboration with your supervisor(s).

A more difficult scenario is where you abandon a topic and turn to a completely different one. The problem then is to determine whether it is best to keep going or take the courageous step of stopping. Recall that there are many examples where a black period is experienced just prior to a major breakthrough — see Chap. 4.

On the other hand, a problem may truly be intractable or, for some other reason, flawed. The difficulty is to be able to recognize whether you should persevere or quit. Here the wisdom of the supervisor can be extremely helpful. However, the student should also be aware that a major change in direction will also be a hard call for the supervisor.

As an example of the value of "riding through the storm", we can quote the case of Diego Carrasco who was stuck on a very difficult mathematical problem. It looked quite impossible. He could easily have given up. However, he suddenly got a lead from the literature and the problem started to fall into place. He ended up with a significant breakthrough due to his perseverance.

An example in the opposite direction was provided by another student, José De Doná. He had achieved some preliminary research results but both José and his supervisor (Graham) felt very dissatisfied with the work. Could it be that the question was wrong? In the midst of this difficult situation, Graham was suddenly asked to write a paper dedicated to the research work of a well-known researcher. Unfortunately, Graham was somewhat unfamiliar with the work of this person at the time. The pressure was huge! Graham asked José if he would mind stepping aside from his research for one week to help Graham understand the new area and to write a joint paper. However, where does one begin? Recall that Graham and José were not familiar with the area. So, to get started, they tried a simple example.

> In engineering research there is nothing like the power of a simple example to throw up new ideas.

As they worked together on the example it became more and more interesting. Eventually they based the invited paper almost entirely on the example. More importantly, the example turned out to be a prelude to a whole new[1] research area in systems and control (Explicit Model Predictive Control). José stopped working on his old project and embraced the new topic. He wrote a wonderful doctoral thesis! Moreover, the thesis was the foundation of a book jointly authored with a colleague — María Seron.

---

[1] Actually several other groups, in different parts of the world, developed similar ideas in parallel.

So, even if you think you have a good question, ...

> Do not be afraid to look around as there may be a better option.

7.6 Changing Supervisor

Changing your supervisor is always a difficult option. One case where it is uncontroversial is when your supervisor suggests it. A less clear-cut situation is when you have a "gut feeling" that your particular case may be better dealt with by a different supervisor. If you are contemplating such a change, then we suggest that you seek advice from other students, colleagues, your potential new supervisor and, in some cases your current supervisor. Be warned that the change will cost you momentum and time. However, whilst we have warned that a swap can cost both momentum and time, it can also accelerate the work if you land in a better situation both scientifically and emotionally.

To encourage readers who may feel that they need to restart their project, it is worthwhile to recount the story of María Seron. María began her doctoral work in the area of passivity and nonlinear adaptive control. This led to three journal papers and four conference papers, more than enough for a doctorate. However, the papers were spread across different topics. She decided to change both supervisor and project and spent time on antiwindup and disturbance rejection in nonlinear control. Finally, in the last 6 months of her program, she worked on fundamental limitations in filtering. The latter topic turned out to be deep and ideally suited to writing a doctoral thesis. Indeed, this latter topic provided the seeds of a book that she wrote with colleagues.

The moral is (Fig. 7.1)

> You can change the topic and complete a doctorate in 6 months if you have the right topic, the right frame of mind and the right momentum.

As with the change of topic, changing the supervisor is a difficult decision so proceed carefully and ethically but, equally, have the courage of your convictions.

Here is pause for thought. A 2005 study claims that roughly a third of employees spend 20 hours a month complaining about their bosses (see Ref. [1], Further Reading). The study was actually aimed at "bosses" to shape up their act. We agree and will return to this in later chapters where we address you as the boss. Here, however, we address you as the student. If you are complaining about your boss (i.e. your supervisor) to the extent that one-third of people do, think again. 20 hours per month, equates to one half of a week every month — just complaining! What else could you do with an extra half week if you spent it more productively.

Fig. 7.1 "You can change topic and complete a doctorate in 6 months if you have the right topic, the right frame of mind and the right momentum", - see Sect. 7.6

7.7 Summary

This chapter considers what is arguably the most important professional relationship you will have during your doctoral studies: interacting with your supervisor.

Clearly, every supervisor has their own style and preferences so make sure you discuss interaction issues with them at an early time.

Section 7.2 emphasizes the importance of co-ordinating the frequency of meetings with your supervisor and of *always* being well-prepared. The frequency of meetings is likely to change during different phases of your work.

The quality of interaction is illustrated by an example we have given of an average, a good, a very good and an excellent student's response to a difficulty.

Finally, in a globalized world, it is also important to be mindful of cultural differences.

Sections 7.3 and 7.4 provide advice on interacting with additional supervisors. For example if you are engaged in an interdisciplinary projects.

The issue of changing topic or supervisor are addressed in Sects. 7.5 and 7.6, respectively. These are difficult situations. You may loose time, or harm a relationship. It is always difficult to know when to persevere and when to change direction.

Sections 7.5 and 7.6 share experiences and pointers to guide you in these, fortunately, rare situations.

7.8 Further Reading

[1] Badbossology Survey, "Complaining About Bad Bosses is a Big Time Drain", www.badbossology.com, 2005.

Chapter 8
The Value of Networking

8.1 Overview

One of the most important aspects of career development (and indeed of life more broadly) is that of networking. This chapter sets out the case for networking and gives some simple guidelines concerning how you can foster and improve this important activity.

8.2 Networking with Peers

There is always great value to be obtained in receiving either technical or moral support from fellow students and other colleagues. Often they face similar challenges and issues. They also have more time than your supervisor. There are certain issues that your supervisor will expect you to resolve by talking to other students before coming to him/her. Examples of issues that have sometimes been inappropriately taken to supervisors include how to use a particular piece of software, or perhaps more commonly, where to get the best car insurance at a reasonable price.

8.3 Other Networks

A crucial aspect of developing a research career is to network with others working in your area of interest. This could include doctoral students at other institutions and, subject to them being open to it, senior researchers at other institutions. As with all networking, be careful not to "overstay your welcome".

The value of networking cannot be overstated. As your career matures your network of contacts will play a significant role in your success or failure. Indeed, most senior researchers, find that about one-third of their total network is comprised of friends and contacts whom they met during their early career.

> The value you will derive from networking in the early years include:
>
> - Exchanging tricks and ideas
> - Helping you solve research problems
> - Moral support
> - Hints on literature
> - Contacts for employment
> - Possible postdoctoral opportunities

> As your career matures, your networks will help you with
>
> - Research collaborations
> - Grant collaborations
> - Sabbatical opportunities

8.4 Attending Conferences

Major opportunities for networking arise at national and international conferences. It is certainly best if you have an accepted paper but, in rare circumstances, your supervisor may be able to fund you to go to a conference when you do not have a paper. This would be especially true if the conference is nearby.

The value of attending conferences is enormous. This is why we included funding to attend conferences as a selection criterion in Chap. 2.

> The advantages of attending conferences include:
>
> - You get to hear the latest results in your field
> - You get to meet the top researchers in one place
> - You get to network with people in your area
> - You get to hear different styles of presentation and learn from them
> - You get to see what others are working on
> - You get to see the "hot topics"

In addition, if you have an accepted paper, then acceptance of your paper calibrates the value of the contribution. This builds confidence and will help you in your final thesis writing. Also, you gain exposure of your ideas and you have the chance to fine tune your presentation skills.

8.5 Joining Professional Associations

One excellent opportunity for networking arises from professional associations. Typically it is quite cheap to join such associations when your are a student. Also, you can use these connections to obtain, or retain, your professional qualifications such as:

- Professional Engineer
- Chartered Engineer
- Chartered Physicist

So we recommend as follows:

> Find, and join, the professional association(s) relevant to your chosen field of study.

8.6 Social Media

In the modern world, there are many exciting new mechanisms for networking. Examples of these media include:

- LinkedIn
- Websites
- Blog/Discussion Forum
- Research Gate
- Google Scholar
- Twitter

The purpose of these media items is to enable people with a common interest to exchange ideas and promote their work. They differ in various aspects e.g. scale, openness, flexibility, etc. All researchers are strongly encouraged to investigate these modern networking opportunities. Some very brief comments on the current media mechanisms are provided below.

LinkedIn: This is a professional social network that allows a user to promote their work. It is widely used as an information exchange mechanism.

Website: This is a way of describing a complete project or an individual's work. Most universities operate a corporate website which contains information about specific projects and academic staff. Often researchers will also set up their own website to give greater exposure to their work.

Research Gate/Google Scholar: These are typically set up by a company (e.g. Google). They have a fixed format and allow others to view your publications.

Blog/Discussion Forum: These are discussion media centred around a person or topic. They create communities. A blog is typically centred on a person, whereas, a discussion forum is typically centred around a topic.

Twitter: This is a networking media which allows one to share information in real time. It can be used to keep up-to-date with others in your field or to share the latest news.

The social media and technology sectors are very rapidly changing landscapes: what is popular today may be forgotten tomorrow. The tools which will be popular in a few years probably do not even exist at the current time. Therefore, the tools mentioned above are just some of the contemporary examples. Remember that the technical methods by which you network, expose our work, collaborate and communicate will change. Yet the importance of these tools will remain! So make sure you are well-informed about the best tools that are currently available.

Clearly, there are advantages in joining or, at least, monitoring these kinds of social media. A simple analogy is that they allow you to attend "virtual conferences" by simply turning on your computer.

At the time of writing this book, Graham has commenced a new research project under the banner "Australian Artificial Pancreas Program". Adopting our own advice (as set out above), we have established a website (www.ArtificialPancreas.com.au) together with associated Blog, Twitter, Facebook and Newsletter connections. Also, we have set up a website (www.ADoctorateAndBeyond.com) associated with the book you are currently reading.

8.7 Summary

This chapter emphasizes the value and importance of networking. We strongly suggest that you hone this skill early in your career because its value will stay and grow with you as you progress. Not surprisingly, networking will be a recurring topic in the remaining parts of the book.

Here are some further issues for you to consider

1. Many senior researchers and managers find that about 30% of their total network comprises contacts that they made during their junior years (see Sect. 8.3). So the contacts you build during the years of your doctoral studies can grow into personally fulfilling and professionally rewarding relationships for decades to come!
2. There is etiquette to networking. Most people have an intuitive understanding of the rules but there are always grey-zones which different personalities and people from different cultural backgrounds might interpret differently. So be sensitive. Some of the corner stones of networking etiquette include:

    - Do not abuse networking - such as pretending to be a personal friend if all you want is some personal gain.
    - There is nothing wrong with being strategic, but remain authentic.
    - Be aware of ethics and social etiquette, particularly in view of cultural differences.

- Be aware of your reputation. If you get known to be a "pushy" networker, a "one-sided" networker that "uses" others, or one with hidden agendas, your reputation will be quickly damaged.

3. There is some finesse needed in networking, particularly when you are approaching a senior person in your field. Being a junior yourself, be sensitive and well prepared. Have your "spontaneous" three-minute pitches ready and clear (see Sect. 11.3). Remember, that senior players are frequently approached by dozens of juniors at gatherings; so be clear, friendly and sensitive to their reactions.

The sections of this chapter covering networking opportunities relevant to doctoral students, including the following:

- Networking with peers (Sect. 8.2).
- Networking with other departments or institutions (Sect. 8.3).
- Conferences (Sect. 8.4).
- Professional associations (Sect. 8.5).
- Social media and technology (Sect. 8.6).

8.8 Further Reading

[1] M. Hubrath "Networking for a successful career in academia" Academics.com
[2] J.A. Stenken, A.M. Zajicek "The importance of asking, mentoring and building networks for academic career success - a personal and social science perspective" Anal Bio Chem, 396:541–546, 2010.

Chapter 9
Tools of the Trade

9.1 Overview

In this chapter we discuss "tools" that are necessary to carry out a doctorate. In particular, we discuss having the courage to keep going when the going gets tough, building experimental equipment, reading the literature and keeping track of references. We also discuss how to achieve an appropriate work-life balance. The latter is a key "tool" whose importance will stay with you throughout your career.

9.2 Have Courage and Tenacity

Doing any challenging task including completing a doctorate, is never a "smooth" journey. The reader may recall the "progress versus time" curve that we presented in Sect. 4.4. The key thing is not to give up the battle.

The following contribution to this book from Brian May (Lead Guitarist of Queen and Astrophysicist - see Sect. 2.2.2) is also incredibly inspiring:

"I think it's about belief... well, about having a clear vision of where you want to be, and believing in it.

Sometimes you have to doggedly press on, in the face of what seems like an impossibility. Sometimes, it seems to me, you have to walk away and come back when opportunity allows. That's what I did. Well, I did both of those things.

All through my studies, both in the 70s and 30 years later, I was in and out of losing belief. I wanted to quit so badly. I think in the end I only hung in because I put my hand up to a friend or two, and asked for help. Help in reading papers, and help in recovering my belief.

I was also lucky I had a wonderful supervisor. If anyone had been able to say that I'd had it easy, because of being famous, or infamous, [my supervisor] would have let himself down, and let me down too. He drove me close to breaking point, but also demonstrated his belief all along. Tough love, I guess! I remember his reaction when I first showed him my transcription into WORD of my whole thesis as written by hand in 1970. It had taken me three months, on and off tour with Queen, 30 years after I'd quit the field of Astronomy to be a musician. I was probably looking for some congratulation, some recognition of my success in getting it all back together, on my own, in the Digital Age. He simply said, 'Have you never heard of Spell Check?'

We went on from there.

As a person who's made a name for himself, in my latter years, I've been awarded many honours, including three honorary degrees, for which I'm very grateful. But these awards never made me feel worthy of adopting the title 'Dr.' The degree that changed everything for me was the one I worked for, in some way, for over 35 years - the one that nearly killed me. For that? Yes, please call me Dr. May!"

9.3 Setting up an Experimental Facility

If you are involved in experimental work, then one of your first tasks will be to set up an appropriate experimental facility. This can be a very time consuming task so we strongly advise that you begin work as early as possible. You will need to source and order equipment, buy software, possibly even manufacture specialized items. The details are problem specific and it is up to you to seek specific help form others on this topic.

9.4 Keeping Track of the Literature

A "tool" that is common to all projects (from highly theoretical to intensely practical) is the requirement of being up-to-date with relevant literature. Important related issues are addressed in the following subsections.

9.4.1 How Much Time Should I Spend Reading the Literature?

The simple answer is as much as you can. A guide would be an average of 30% of your time but, in the beginning the percentage will necessarily be higher.

The advantages of reading the literature include:

- Learning new techniques
- Becoming familiar with the field
- Getting new ideas
- Avoiding the "reinventing the wheel" syndrome

The only way to stay current is: to stay abreast of the evolving literature. Also, remember that you cannot hope to publish a paper or have your thesis accepted if you are not totally familiar with the appropriate published literature.

After investing a lot of time in reading the literature, you need to be skilled at managing your evolving bibliography. This brings us to our next topic.

9.4.2 Keeping Track of References

One of the key tools required to carry out research is managing the literature and references that you collect during your studies. A typical doctorate thesis might include some 200 references that you collect over a two to five year period. It is therefore essential that you have a disciplined approach from the very beginning of your work. You need to store the papers in an appropriate electronic format and have a suitable digital referencing system. It is very important to start early.

While reading the literature, it is good practise to summarize each paper you read. You should consistently write down what the relevant results are, and any insights you might have gained. This should be a habit developed from the very beginning as it will definitely prove useful later down the doctoral road. Having the summaries available will simplify the task of making the correct reference when writing papers and writing your thesis.

Even more importantly, it is a key prerequisite when presenting the literature review of the topic your work is based on. It is very easy to look past this simple advice. One of the worst feelings is to realize you have to spend several weeks re-reading all the papers you read during your studies when you are at the end of your doctoral studies (and probably short of time).

Many different software tools are available to help you streamline the process of finding, downloading, storing, organizing, categorizing and annotating the papers or documents you come across. Some of these tools are Papers®, Mendeley®, Readcube®. One of the advantages of using these applications is that they also allow the user to export bibliography databases in many formats. This means that, if you are already using one of these applications to manage your papers, then simply by selecting the ones needed, an automatic database can be exported with a couple of clicks of the mouse. This saves a lot of time compared to creating the database manually. The difference is that the effort and trouble is shifted to the beginning of the timeline, when you add the paper and add or correct all the metadata. As with social media, there are new tools appearing all of the time - so make sure you know what is on the market and what your peers are using. We provide a specific example of a tool currently in use by some researchers for reference tracking on the website for this book.

9.5 Other Tools

Beyond referencing you will also need to master other tools that are applicable to your particular field, e.g.,

- Wordprocessing tools (e.g., Word® or LaTeX®)
- Mathematical and computational tools (e.g. Mathematica® or MATLAB®)
- CAD tools (e.g. Autocad®)
- Presentation tools (e.g., Powerpoint® or LaTeX®)

For example, in the area in which the two authors of this book work, great proficiency in LaTeX® and MATLAB® is an absolute prerequisite. In other areas, these tools are rarely, if ever, used. So make sure that you are aware of the tools that are in vogue and accepted in your specific area.

Note that you can, and should, use your network to learn and adopt the best practices for the tools you need.

9.6 Work-Life Balance

> One of the "tools-of-the-trade" which will remain relevant throughout your career is that of maintaining the right work-life balance.

Indeed, we devote an entire chapter to this topic in Part IV of the book. (See Chap. 23.) In particular, we point the reader to Sect. 23.4 ("Working Smarter not Harder") and to Sect. 23.5 which discusses approaching time management from an energy management point of view. The concept is equally valid when you are doing your doctorate as it is in later life.

The following advice was provided by Diego Carrasco (one of Graham's former doctoral students):

"It is fairly easy to fall into the routine of going to work at university, coming back home, sleeping and repeating the sequence over and over. This is especially true if you are single, living alone and in a new country.

The first shock comes as you realize that you have just left all your known life behind: your friends, your family and even that local bar you used to frequent. Nothing you knew is there anymore, and you only have yourself to rely on.

Some of you may be more sociable than I was, but it was easy to fall back in the routine of work/home/sleep, I did not question it. The main reason was that I was swamped with things to do for work, or so I wanted to think. Do not get me wrong, I had friends, I went out for drinks, we had barbecues on the weekends, etc. But I did not put special effort into creating new situations or into meeting new people. I relied on the already existing network of other doctoral students in my department, all international students in a similar situation. I met great people. But there is a catch: it is a closed environment and, as I found out later, everyone moves on. Many of my best friends are now scattered all over the world.

One day, a friend of mine said something along the lines, "If you plan to stay in academia and do this for the rest of your life, you have to live your life in a way that you will be comfortable with for the rest of your life. You have to start shaping that now, otherwise you will never do it". Then it hit me. Yes, I wanted to do this for the rest of my life, but no, I was not happy with my routine. What kind of life would I have being at work for more than 10 h a day, sometimes even on weekends? My social life was reduced to activities other people organized, I was just going with the flow.

So I took steps towards improving. I moved in with two other friends. I joined the university soccer club, I started exercising regularly, and later on I even took dancing lessons. All of these steps were great decisions. I still spend a lot of time at the office, but I am now much happier with the way I live and the way I deal with stress. I make the effort to meet new people. This has given me the opportunity to meet amazing people that have had influence on my life in many ways."

The take-home message here is this:

> Take life by the horns, decide what you want to do and do it. No-one is going to do it for you. Take this as a chance to know your weaknesses and to improve yourself.
>
> Be efficient managing your time, both in and outside of work, do not slack off and settle for anything less. Do what makes you happy. Make the effort. It is worth it.

9.7 A Long-Term Vision

We will devote a substantial part of a later chapter to this topic (see Chap. 24). However, it is important to note that, if you are going to have a long-term vision, then the time to start working on it is now! For example, (Fig. 9.1)

> You may want to produce such high quality research that you ultimately become an FRS. For some the letters FRS will stand for Fellow of the Royal Society, London. For others, the letters FRS might have an equally strong appeal but with an entirely different meaning.

9.8 Summary

This chapter discusses two categories of tools: first, specific technical tools (often computer-based) and, second, organizational skills such as collecting bibliographical references during your thesis work. We also introduce the topic of achieving the right work-life balance.

Fig. 9.1 "You may want to produce such high quality research that you ultimately become an FRS. For some the letters FRS will stand for Fellow of the Royal Society, London. For others, the letters FRS might have an equally strong appeal but with an entirely different meaning".- see Sect. 9.7

9.8 Summary

The technical tools are quickly summarized: you should check what is currently available and best-practice. New options are evolving so fast, that an in-depth coverage in a book like this would make it quickly obsolete. Our point is to familiarize yourself with whatever tools you need; become a whiz! The more the use of these tools becomes second nature to you, the more you can concentrate on "what" you are using them for - your research results.

Specific technical tools comprise many things, e.g. word processors, CAD, mathematical and technical software and hardware. Familiarize yourself with what is being used in your department and use your network of students in your own and other departments and universities to see what tools are considered best-practice in your field.

Turning to organizational skills, Sect. 9.4 singles out the important issue of collecting your references and bibliography as you go along. This task also involves a software component, since there are several software applications available that help you to accomplish this task. We mention one such tool on the book's website but you should use whatever you are comfortable with, preferably something that others use in your department.

Independent of the tool you use, we emphasize the importance of doing so systematically and from the beginning of your doctoral studies. Do not fall into the trap of thinking you can relocate hundreds of references in the literature at the end of your thesis work (see Sect. 9.4.2)!

The other skill we address is that of work-life balance. Like networking, this is a life skill that will accompany you throughout the remainder of your career.

During your doctoral, work-life balance issues can arise in several ways:

- The rigours of full-time research and the stress resulting from your research journey can place a significant burden on health, family and social life.
- In the case of part-time research, you have the added challenge of being torn between the different strains of research and work.
- If you are studying in a town away from your home, or even country, you can become socially isolated.

Our advice is not to be fearful of these challenges, but to keep an eye on them. We give initial insights on how to do so in Sect. 9.6, where we also provide forward references to later chapters.

9.9 Further Reading

[1] P. Gosling and B. Noordam, "Mastering your Ph.D.: Chap. 13 Searching the scientific literature," Springer, 2nd edition, 2011.

Chapter 10
The Art of Publication

10.1 Overview

The fact that research necessarily contains innovative and original ideas, invariably means that the ideas should be publishable. Indeed, publication, in an appropriate peer-reviewed scientific conference or journal, is the ultimate proof of the validity of your work. Therefore, the step from doing research to publishing is inescapable. The only time that one might not consider publishing is when there are intellectual property (IP) or other classified restrictions on your work. However, such restrictions usually hold for a finite time window, e.g. until a provisional patent is received. In such a case, publication is still ultimately of great importance.

10.2 Publications

There are definite advantages to obtaining publications in a refereed conference or journal during your doctoral studies.

> Advantages of writing papers include:
> - Building confidence.
> - Establishing additional evidence that you are generating innovative results.
> - Giving exposure to your work.
> - Helping generate a strong Curriculum Vitae which can be crucial in gaining employment after completing your thesis.
> - Providing a catalyst for networking.
> - Aiding in the final thesis write-up.

In particular, in the author's experience, having publications "along the way" greatly facilitates the writing of the final thesis. (See Chap. 13 for guidelines relating to "How to write your Thesis".)

Of course, publication rates and expectations depend upon the nature of the project. In some cases, progress can be measured by the completion of sub-tasks whilst, in other cases, the project reaches its climax in one final sweep. The latter is especially true in projects involving hardware. In all cases,

> Writing papers is a core part of research and is an essential ingredient in an academic career.

10.3 How Many Papers Should I Write?

10.3.1 Conference and Journal Papers

The desirable number of conference or journal papers depends on the topic and, to some extent, on the field. In the authors' experience having three or four conference papers and one or two journal papers greatly aids the writing of the final thesis. Of course, this is arguably simply a confidence trick, since instead of insisting that students climb one high mountain (the thesis), he/she is being asked to climb four or five other high mountains (the papers). Nonetheless a serious aspect of writing papers along the way is that the work load is distributed rather than being concentrated at the end. Also, one gains all the advantages listed in the previous section.

10.3.2 Is It Possible that a Book Arises from My Thesis?

There are many examples of this happening. Indeed, over 10% of Graham's 40 doctoral students transformed and embellished their thesis into the form of books. If this happens, it can be a great start to your career. It has all of the associated advantages listed in Sect. 10.2 but with even greater intensity! However, one should not view writing a book as a necessary outcome but simply as a nice extra step if it occurs.

10.4 Writing a Good Paper

Having established that you need to write papers, the next logical question is, how does one write a *good* paper? The first and critical aspect is to have a publishable idea. This should be decided in collaboration with your supervisor. Then,

> Take time writing the paper! A great idea can be destroyed by a poorly written paper.

In the authors' experience, junior students tend to treat paper writing somewhat like constructing a smorgasbord. They put every idea down in a somewhat random array and hope their readers (and reviewers) will see the great ideas "on the table". Alas, reviewers tend to be busy people and may therefore overlook the key aspects of your papers. Remember, it is up to you, the author, to sell the ideas.

We believe that writing a great paper has much in common with writing a good research proposal (see Sect. 15.3.3).

You should:

> - Carefully decide what it is you want to claim in the paper.
> - Say how others have approached the problem.
> - Clearly state your "killer idea".
> - Prove, or illustrate, the idea as clearly as possible.
> - Summarize the result clearly.

> In some cases, writing a great paper can take almost as much time as doing the research in the first place!

One trick that the authors have cultivated over the years is to apply the three-minute rule (see Sect. 11.3).

> Ask a colleague (or preferably a joint author) to sit with you while you explain the concept of the paper in three minutes.

In three minutes you should be able to say exactly what the paper is about. This will create a template for the final paper.

We have also found it useful to write papers in a top-down fashion, i.e. begin with the abstract, then the set of section headings, then a two line description of the section and finally the details.

10.5 Peer Review and Resilience

One of the key features of publication is that the work will be subject to peer review.

Peer review has many advantages, e.g.
- Calibrating your work.
- Pointing out unknown links to other work.
- Providing advise on the best way to "tell the story".

However, at times, reviews of publications can be very upsetting. Reviewers can simply miss the point of the paper. Worse still, they can have their own agenda. Our strong advice is that you approach the review process with a positive attitude.

In the event of a rejection, you have two options:
  A: Take the rejection negatively and become angry.
  B: Take the rejection positively and proceed to do one of the following:
- Rewrite the paper responding to the reviewers criticisms. Then resubmit the paper to the same venue with a detailed commentary describing how the paper has been modified.
- Rewrite the paper and send to another venue.
- "Lick your wounds" and begin working on your next, even better, paper.

Obviously, we recommend approach B.

Graham vividly recalls an incident that occurred to him and two co-authors during their early careers. They believed they had solved a long-standing open research problem (actually on the topic of convergence of Adaptive Control Algorithms). They submitted a paper to one of the top journals in their area.

Soon the reviews of the paper arrived. By way of background, we note that the journal, to which we submitted the paper, asked reviewers to rank papers on a four-point scale:

1. Excellent — (publish as it is)
2. Good — (publish with minor corrections)

10.5 Peer Review and Resilience

3. Promising — (might be publishable after major corrections)
4. Wrong — (never publishable)

We had four reviewers and guess what? Yes, we got four reviews at level 4. That's what one calls a rejection!

However, we felt strongly that we had a valuable contribution. So, we did a careful analysis of the reviewers' comments and wrote a firm but polite rebuttal.

Fig. 10.1 "A rejection isn't always a bad thing" - see Sect. 10.5

The paper was accepted! Moreover, the paper was ultimately selected by an expert panel to be one of the 25 most influential papers in Control Theory published between 1932 and 1981.

So now, when our own students come to us with their heads between their legs saying, "They rejected my paper!", we tell them the above story. We conclude by saying (Fig. 10.1)

> A rejection is not always a bad thing.

This brings us to the important topic of resilience. This is one of the key life skills that will stand you in good stead through your life. Unfortunately you will encounter many set-backs as your life evolves. They will be of both a personal and professional nature. How you respond to these set-backs will define you as a person and have a major influence on your ultimate success. You could become angry, cynical, depressed, anxious, etc. These steps will get you nowhere fast. However, if you can bounce back with enthusiasm and a fresh resolve to succeed, then you will find the success, you seek. For more on this important topic we also refer to Sect. 24.3 "Regaining Balance when Injustice Strikes".

10.6 Writing Patents

For those working on applied projects, the possibility of writing one or more patents arises. Some industries may even insist upon this as part of the collaboration process.

> Patents can be extremely beneficial in securing the Intellectual Property of your work for commercialization purposes.

> For an invention to be patentable it must be novel, it must involve an inventive step and it must have a useful application.

Patent law can be complex and you should seek professional advice. For example, the US Patent and Trademark Office recognizes six types of patents: Utility Patent, Design Patent, Plant Patent, Reissue Patent, Defensive Publication, Statutory Invention Registration.

There are also several steps in the patenting process. For example:

- A Provisional Application: This step is relatively inexpensive and allows you to obtain an early priority date on you invention. It is not a full "patent" but it will

provide you with a time window to decide whether your idea is worthy of filing a full patent.
- A Full Patent: This gives you protection of your invention over a period (typically 20 years).
- Patent Cooperation Treaty (PCT): If you file under a PCT then you can seek protection in up to 148 countries in the world. Note that you need to pay a fee covering each country in which you wish the patent to hold.

If you believe you have a patentable idea, then you should talk to your supervisor and/or industrial collaborators. Note that there are strict rules on who should be named as inventor on a patent. You can obtain appropriate advice from the commercial wing of your University. They will typically work with you to decide if the work is patentable. They may also provide funding support if they are enthusiastic about the idea you wish to patent.

Note, that writing a patent is very different to writing a paper. Writing a patent requires a special language and focus. We suggest you use other patents available in your area as a template. Typically, the inventors write a rough draft and then seek help from a professional patent attorney who has specific expertise in this area. Obviously costs are involved. However, a good patent attorney can make the difference between success and failure.

10.7 Summary

The ultimate proof that you are engaged in innovative research is getting your work accepted in a peer-reviewed conference or journal.

There are several advantages to be gained from publishing your work, even before you have completed your thesis: it builds confidence, it creates "modules" of material that you can use in your thesis, and it allows you to network at conferences (see Sect. 10.2). Make sure you liaise with your supervisor and seek advice.

The peer-reviewed publishing process consists of several steps, e.g.

- writing,
- submitting for peer review and
- rewriting to address reviewer's comments.

The above sequence is always traversed at least once but usually two, three and possibly even more times.

As far as writing is concerned, our key advice is to take your time — the value of a new research contribution can be severely undermined and obscured by a poorly written paper! After devoting so much time to developing the idea addressed in your writing, do not short-change yourself by allocating inadequate effort to *communicating* the idea.

As you read the literature relevant to your doctoral studies, take some time out to study *how* a good paper is written (as opposed to the content): how is it structured;

how does it capture (or fail to capture) your interest; how does it embed the new results into the known ones. Learn from the literature you read, hone your writing skills, see Sect. 10.4.

The second step of the iterative publication process is to receive and integrate the reviewer's comments. It is easy to become defensive when something you have devoted so much effort to is judged, criticized and challenged. This is true for senior authors, and even more true for your first publications!

Breathe deeply, and be open (Sect. 10.5). Walk away from the reviews for a day or two if that helps you gain perspective. Then come back, build on them and make a better paper.

Try to carefully understand the type of feedback you have received: Is there an actual error? Is there a contradiction to previous results in the literature that needs to be addressed? Are the conclusions poorly substantiated? Have you failed to put your work into a proper perspective with existing literature? Do the reviewers agree with the content but feel it is poorly communicated?

Make review feedback *happen for you — instead of to you*!

We assure the reader that this book has become a better book due to the reviewer's comments we received (see Acknowledgements).

In a few cases (maybe some 10%) the insights created during thesis work will become the catalyst for a book or a patent, but we recommend that you see this as a bonus if it arises naturally, rather than to obsess about it (Sects. 10.3.2 and 10.6).

10.8 Further Reading

[1] G. Gopen, L. Swan "The science of scientific writing" American Scientist, November 1990.
[2] J. Schimel "Writing Science: How to write papers that get cited and proposals that get funded" Oxford University Press, 2011.
[3] R.A. Day, B. Gastel "How to write and publish a scientific paper" 7th edition Cambridge University Press, Cambridge 2012.

Chapter 11
The Art of Making Great Presentations

11.1 Overview

Throughout your years as a student, and in your later career, you will need to make presentations. The audience you address may range from small to large groups, from experts and students to lay people. Knowing who is in your audience, and what their backgrounds are, is an essential first step. Your ability to make a great presentation will play a huge role in your ultimate success. Thus, we strongly suggest that you practice the art of making presentations. Some brief guidelines are given in this chapter.

11.2 Conference Presentations

In Sect. 8.4, we outlined the advantages of attending conferences. Of course, all of these advantages can be lost if one gives a poor presentation.

The first step is to have something that is worth talking about. However, this is by no means sufficient. A poor presentation can ruin very worthwhile topics, even undermine your credibility for future talks.

A preliminary guide to giving a good presentation is provided by Toastmasters (see Ref. [3], Further Reading) who recommend progressing in three stages (Fig. 11.1):

1. Tell what you will say.
2. Then say it.
3. Then tell what you have said.

Whilst this may sound redundant to less experienced speakers, this is a powerful technique. The rational is that in the first step you get everybody on the "same page" by setting the scene. Then, in the body of the talk, you deliver the substance of your talk. Finally, you conclude the talk by summarizing the essential "take-home message".

Fig. 11.1 "1. Tell what you will say. 2. Then say it. 3. Then tell what you have said." - see Sect. 11.2

Another way to look at why this approach works so well is as follows. The substance of the presentation (middle part, "Say it") is framed by first capturing the attention of an audience that has not heard it yet ("Say what you will say") and, in the end, imprinting the message on the now informed audience ("Say what you have said": see also Chap. 20).

Other aspects of a good talk include:

- It should be succinct.
- It should be relevant.
- It must be appropriately titled.
- It must be enthusiastically delivered.

To expand on the last point, you should

11.2 Conference Presentations

- Speak "to" and not "at" the audience.
- Project your voice to the whole audience. (This is a skill that can be practiced and fine tuned.)
- Speak slowly—bearing in mind that some members of the audience may not have your language as their native tongue.
- Have clear and simple visuals (avoid tiny print, too many formulae, etc.).
- Remember that an appropriate pause can be more powerful than many hundreds of words.
- Watch your time budget.
- Talk about something the audience is interested in rather than what interests you.

A good talk also avoids the following mistakes:
- Rambling
- Boring
- Too technical
- Not audible
- Contains apologies
- Confuses appropriate humility with pity seeking claims of your inadequacy hoping for an underdog advantage. It usually backfires.

It is always a good idea to try a "dry run" of your talk with a colleague. Also, accept any criticisms they give as a source of learning rather than becoming defensive or upset. Recall what we said about reviews on Sect. 10.7: make feedback happen *for* not *to* you.

Note that it is possible, maybe even highly likely, that members of your audience will be bored or simply uninterested. You need to engage them in what you are saying so that they *want* to listen!

Finally, you will know when you have delivered a great talk if members of the audience follow up with questions or seek to engage you in conversation after your talk.

We cover other points on the topic of public speaking in Chap. 20 and suggest it could be beneficial to quickly read that section at this point in time.

11.3 Three Minutes Is All You Have

This is a powerful technique where you get three minutes only to explain to someone else a whole topic such as your thesis, project, paper or book. The point is not to speak fast but to be succinct and to focus your mind. If you are able to get your complex topic across in three minutes then you know you have truly mastered it.

Do not be misled into thinking this is an easy task (after all it's only three minutes). Instead it is a demonstration of great depth of knowledge. Only deep understanding will allow you to convey the essence of a complex topic in just three minutes.

Typically, the background preparation for the three minutes can absorb several days of work. It is definitely a skill that requires practice!

Also, try to use this technique with different types of listeners such as your supervisor, nontechnical friend or a person from industry; each time adapting your message and language to the target audience.

One useful trick is to try this before you go to a conference so that you can succinctly get your message across whilst you are networking. For example, when somebody approaches you and asks "what are you working on?" or "what is your paper about?", then you will have a clear and succinct answer. Some people call this "the elevator pitch", i.e. you meet your boss in the elevator and you have just three minutes to convince him/her of some idea. This is a skill that is definitely worth perfecting!

Finally, the three-minute technique allows you to verify whether or not the topic you are working on has sufficient substance. We remind the reader of the example given in Sect. 3.4 of the student who actually concluded that his three-minute description of his research was impossible because he did not have the depth or quality of results to talk about it at this intense level.

We want to emphasize that we are not criticizing this student. Instead, we highly commend him for his honesty in realizing that, when asked to distill the work he had done in three minutes of pure essence it was: vacuous. Whilst this may, at first, seem shocking it was much better to discover this than to proceed in ignorance. The key point is that your three minutes are precious and should not be filled with hollow ideas.

11.4 Other Ideas

Many books and papers have been written on the art of making good presentations. We strongly suggest that you read the literature and practice your skills in this area. We also present other ideas in later chapters—in particular see Chap. 20.

11.5 Summary

The companion skill to writing about your work is being able to talk about your work. Just as a poorly written paper can obscure a great research result, a poorly delivered presentation can also undermine understanding and interest in your work. So take time to prepare your presentations properly—the audience does not see the years of effort you have spent researching your result, they only experience your delivery of it!

Presentations can range from being as short as three minutes, to the typical conference presentation length of 10–20 min, or longer.

No matter how long your presentation, we always recommend being able to present the gist of it in three minutes (Sect. 11.3). You will know when you have truly mastered your complex subject if you can succinctly communicate it in just three minutes.

Your three-minute presentation (or elevator pitch) is also an excellent basis to then embellish into the complete presentation.

Section 11.2 gives advice on how to structure, support and deliver your presentation. In particular, Toastmaster's recommended structure described in the section is well tried and tested.

Finally, a great way to become a good presenter is to watch others. Learn what works and what does not. There are also plenty of great presentations to watch on the Internet, such as the well-known TED talks, which are similar in length to a typical conference presentation.

11.6 Further Reading

[1] P. Gosling, B. Noordam "Mastering your Ph.D.: Chap. 12 Mastering presentations and group meetings" Springer, 2nd edition, 2011.
[2] C. Gallo, "The presentation secrets of Steve Jobs" McGraw Hill, 2010.
[3] Public Speaking Tips - Toastmasters International

Chapter 12
The Ethics of Research

12.1 Overview

The Ethical conduct of research is one of the key issues of which *every* researcher should be aware. The authors know of several students whose doctoral degrees were denied because of a breach of ethical standards (e.g., submitting the same material simultaneously to two separate institutions, or plagiarism).

In recent times, several well-known academics, politicians and business people have lost their jobs and destroyed their careers by failure to comply with ethical standards.

An internet search on this topic is quite enlightening. It will reveal a large group of high-profile and promising careers that were destroyed due to academic fraud, often motivated by something as mundane as vanity. The current chapter is intended to make you aware of some of the associated issues.

12.2 The Context

> Every researcher must understand the code of research ethics applicable within which they work.

Universities and research institutions all have detailed codes of ethics. This chapter combines ideas from several of these.

In the sections presented below we cover some of the basics of conducting ethical research as a doctoral student. In a later chapter, we will expand on these ideas to cover broader issues. You should check the guidelines from your own institution to see if there are other important issues not addressed here.

12.3 Humans, Animals and the Environment

In many areas of scientific research, there are potential physical or emotional risks to humans or animals. In these cases it is important that you understand the potential danger:

> Under no circumstances should research pose a risk to the safety of humans, animals or the environment.

Also,

> It is essential that prior approval be sought from the appropriate ethics committee covering research involving humans or animals.

Failure to do this is not only unethical but could result in criminal charges.

12.4 Keeping Accurate Records

Researchers are responsible for the accurate recording of data.

> All data needs to be filed in an appropriate fashion so that the veracity and timing of results can be verified by an independent authority.

Failure to maintain appropriate records could lead to severe consequences. Indeed, the authors have witnessed several law suits which revolved around the timing and accuracy of research results. The penalties can be extreme including loss of your job. Even more importantly you could loose your reputation and self-esteem.

12.5 Accuracy of Publications

> All publications should be accurate and honest.

12.5 Accuracy of Publications

> Under no circumstances should results be presented in such a way as to obscure their true meaning, impact or source.

12.6 Plagiarism

12.6.1 Citing the Work of Others

Extensive inclusion of other people's work without proper referencing amounts to what is know as "plagiarism". Violation of this principal is unethical, a violation of copyright, and a violation of various professional codes. We encourage you to do an online search of famous people that were caught with plagiarism: besides the obvious damage to reputation, entire careers have been destroyed! Thus

> All researchers need to accurately cite the work of others.

Failure to cite prior work properly is a sure fire way of getting your own work rejected and your career destroyed, so be warned!

12.6.2 Citing Your Own Work

Citing previous work does not only refer to other people's work: it also refers to your own previous work! Failure to do so is labelled by the slightly confusing term "self-plagiarism". It might seem confusing at first because, how can you plagiarise ("steal") something that already belongs to you?

On second thought, however, and with a little research you will find that the issue with this offense is not due to violating your own copyright but the secret inflation of your contributions and of your number of publications. Thus, if you publish the same result in a slightly modified form in different journals it would appear that your number of publications is higher than is warranted.

The fact that you have not referenced your own work is interpreted as a deliberate and major offense. This is perfectly reasonable. After all, since it is your own work, you definitely cannot claim to have been unaware of the work.

To understand how seriously this offense is taken, we again encourage the reader to do some on-line research on this topic.

In a famous case in Germany, a young professor who had won many awards, held a prestigious position and had been called a "rising star" was caught having published over a dozen similar papers without ever referencing himself—the verdict was clear: self-plagiarism. As a consequence he lost it all: the publications were retracted, the awards taken from him, the job lost—and, once again, career and reputation, were damaged beyond repair.

12.7 Multiple Submission of Papers

Related to the previous section's topic of self-plagiarism is the possible temptation to submit a paper, possibly in slightly modified form, to several journals or conferences simultaneously. Clearly, if they were all published, this would amount to the offense of self-plagiarism.

A young researcher might think they could do this purely to increase their chances of being accepted in at least one of the journals or conferences. They might also be tempted to think that this is not too bad as they intend to withdraw the other submissions once one of them is accepted. However:

> Under no circumstances should multiple submissions be contemplated.

All reputable journals monitor this closely. Submitting multiple submissions often has severe penalties including being banned for a substantial period from publishing in that journal! Moreover, the indiscretion often becomes widely known and can undermine an individual's entire career. The authors know of several cases where this has actually happened.

12.8 Who Should be an Author of Publications?

> It is important that all people who contributed in a substantial way to a piece of research be included as authors, or at least be acknowledged in the associated publications.

The question of authorship needs to be discussed openly and an agreement reached. This includes the order of authors. You need to decide whether the order should be alphabetical (to reflect even contributions) or carefully arranged (to reflect different amounts of contribution).

> All authors need to agree on an order in which the authors will appear.

Also, in the case of (near) equal contributions to multiple papers, it may be desirable to rotate authors so that "Zarrop" gets an equal number of first places as does "Anderson".

12.9 Summary

Ethics is an important topic that everybody should internalize from the beginning. The chapter discusses ethical conduct in three broad categories.

First, there is the code of ethics applicable to your university, institution and possibly even your field of study (Sect. 12.2). These codes usually include ethics of conduct, research that affects life and the environment as well as obtaining required permissions. It is your responsibility to familiarize yourself with the relevant codes.

A second broad category affects honesty and openness in research (Sects. 12.4 and 12.5), including record keeping and honesty in publication.

The third category covers various forms of plagiarism. Section 12.6.1 covers the most commonly known form of plagiarism, which refers to copying or building on the work of others without proper referencing. Section 12.6.2 covers the equally serious, but less well-known offense, of self-plagiarism. This is the illegitimate attempt to inflate one's credibility or number of publications by submitting the same results repeatedly and concealing this by not referencing your own (repetitive) work. But even when you reference yourself, multiple submissions are not permitted (Sect. 12.7).

Finally, ethical behaviour extends beyond codes and laws. It also includes ethical behaviour towards your fellow students and colleagues. An example of this is deciding on the authors of a paper and on the order of their names on publications (Sect. 12.8).

The chapter encourages the reader to take the topic of ethics very seriously. Even unintended violations can have serious consequences—if one is lucky, these consequences become clear at an early stage and you can immediately take corrective action. In the worst case, they appear much later when the consequences are much harder to correct. We recall that many, senior politicians and researchers have lost everything years down the line when much more was at stake for them.

Chapter 13
How to Write Your Thesis

13.1 Overview

By this stage, we assume that you have all the necessary components in place to embark on the closing task of completing the actual writing. Several books (see, for example Reference [1], Further Reading) have been written on this general theme. We will focus on the following topics.

- When to start writing.
- Planning.
- Reviewing.
- Length.
- Being self-critical.
- The alternative of gaining a doctorate through publications.

13.2 When to Start Writing

Obviously, the formal writing process occurs when you believe you have brought the work to a reasonable conclusion. There are several things that will greatly aid the writing process, including:

- If you have been diligent about writing papers along the way, then these can act as a first draft of key chapters.
- If you have had the final write-up in the forefront of your thinking as work has evolved.

Indeed, we think it is a good idea to have the final thesis in mind before setting out on the doctoral journey since, ultimately, you will have to put the ideas into your thesis.

13.3 Planning

As in all jobs, it is a good idea to carefully plan the work before you write a single word.

> When writing the final doctorate, we recommend that you first write the abstract to get the "story line" correct. Then assemble your papers and reports into chapters so that the story flows from chapter to chapter.

Also, you definitely do not need to be constrained by the sequence in which the work was actually done. As we have said elsewhere, the progress of doctoral work can be a messy business. However, by the time you begin writing up, the whole story should be clearer. Thus we recommend:

> Do not be constrained by the sequence in which the work was done or by the sequence of the associated papers, but choose an order that "tells the story well".

Provided the papers and reports are available, then the above "approach" need not be overwhelming.

The main thing is to have a clear vision of what the research is about. You should make sure that you read Sect. 11.3 (3 min is all you have!).

> The first step in writing a thesis is to first answer four questions:
> 1. What exactly have I done?
> 2. Why is it novel and interesting?
> 3. Why should others be interested?
> 4. How can I best tell the story?

13.4 Reviewing

Writing a thesis usually requires multiple phases. Thus write a first draft relatively quickly, then review and rearrange to improve the story. Get feedback from your supervisor. Revise and build you thesis. However, be careful to avoid "mission creep". Try to settle on a good layout as soon as you can so that the final thesis begins to take its final shape at an early stage.

13.5 The Target Length

The question of an appropriate length for a thesis is discipline dependent. In applied areas, it may be desirable to give full details of the experimental apparatus and techniques. In more theoretical areas, it may be possible to give a tight and succinct presentation.

Some of Graham's best doctorate students wrote theses comprising between 150 and 180 pages. Others required more, (up to 600 pages). On balance, we have a preference for tight, elegant theses but we re-iterate that this is specific to different disciplines.

13.6 Be Self-Critical

You should obtain detailed feedback on your thesis draft from your supervisor - indeed that is one of their key roles. Try not to be defensive about negative comments. Accept them in the spirit they are delivered, i.e. in an effort to best help you. Remember:

> Make feedback happen *for* you and your growth.

13.7 The Alternative Route of Presenting Papers Only

Many universities are now allowing doctorate students to submit a portfolio of published work instead of writing a formal thesis. This route is often adopted by more senior people who have amassed a substantial set of publications and want it recognized as being equivalent to a regular doctorate.

For most students, there is not much difference between this route and the route of writing a formal thesis. After all, the first draft of the chapters in the thesis is likely to be the author's papers. Beyond that, the thesis allows the student to "retell the story" with the benefit of hindsight using a consistent notation etc.

13.8 Summary

This final chapter of Part II of the book has addressed the process of actually writing your doctoral thesis. This includes aspects such as when to start writing (Sect. 13.2), the importance of planning (Sect. 13.3), reviewing the content, length and being self-critical (Sects. 13.4–13.6).

Asking yourself the following questions is helpful when outlining your thesis:

1. What exactly have I done?
2. Why is it novel and interesting?
3. Why should others be interested?
4. How can I best tell the story?

When it comes to the fourth point, it is important to detach yourself from the sequence in which you did the actual work. The thesis is not a historical account, it is an exposition of research and results that should be relevant and interesting to the reader. It should build a pillar for subsequent work, be it by others or by yourself.

13.9 Further Reading

[1] P. Gosling, B. Noordam "Mastering your PhD: Chap. 19 Writing your doctoral thesis with style" Springer, 2nd edition, 2011.

Summary of Part II

This part of the book has provided a guide to an individual carrying out doctoral studies.

The first two chapters address student-supervisor interactions and the benefits of networking. The next three chapters describe tools of the trade, the art of publishing and the art of public speaking. We have also made the reader aware of the ethical guidelines within which all research should be conducted. Finally, we have given advice on the all important task of actually writing the final thesis.

Part III
Using Your Doctorate: The Early Years

Overview of Part III

Part III of the book moves onto the period beyond your doctoral studies. However, it may also be helpful to read this part before commencing your doctoral journey since it will further inform you of issues and challenges that you are likely to meet after you graduate. We address such issues as securing a job (Chap. 14), life as a young academic (Chap. 15), life in industry (Chap. 16) and moving between academia and industry (Chap. 17). We also cover the evolving cycle of success (Chap. 18).

Chapter 14
Securing a Job

14.1 Overview

After completing your doctoral studies, the next phase of your career is likely to kick-off with seeking a suitable position in either the academic, government or the private sector.

14.2 Where to Apply

We first compared the different sectors in broad terms, details will be covered in later chapters. We then raise the question of what level to apply for; in particular, two options include to either aim high or to get a strategic "foot in the door" and work yourself upward from there. The final sections of the chapter cover pointers for a successful job application and the all-important job interview.

In university positions, the emphasis is on teaching and on individually chosen research projects. Industry positions are likely to be more target driven with a greater emphasis on being entrepreneurial.

Both academic and industrial positions involve some level of administrative responsibility and the need for community interaction.

By "community" we mean affiliated groups and organisations. In academia these include members of your department, other faculties, other universities, industrial collaborators, professional organisations, etc. In the private sector, they include other staff and units in your company, other companies, academic collaborators, suppliers, clients, professional organisations, unions, etc. Government positions lie somewhat in the middle. For more detail help to your decision-making see Chaps. 15–17 where we cover specifics of a job in the academic sector, private sector, as well as moving between sectors.

> It is always helpful to ask people that you know in each of the sectors what their job is like. This is another situation where the network you built during your doctoral studies will prove extremely useful.

14.3 Selecting an Appropriate Level

Having decided on the sector, another question is the degree of challenge associated with the job that you seek. It is generally accepted that being under-challenged leads to poor performance due to boredom and lack of stimulation, whereas being over-challenged leads to poor performance due to stress and burn-out. Thus you need to position yourself carefully.

All else being equal, we believe that you should aim as high as you possibly can, since high achievement rarely occurs if you are 100% within your comfort zone.

As many others have experienced, for both authors of this book, the stretch positions that they accepted were defining moments in our careers. So aim wisely!

As one of our friends (an ultra marathon runner) said,

> "Success depends on being courageous but not so much that it leads to injuries that set you back".

Of course, you must not just blindly aim high and pray; if your current approach is consistently not bearing fruit you might have to revise your strategy.

Sometimes, there may also be external factors that either force you, or at least make it prudent, to choose a lower entry job and work your way up. Such factors could include economic recessions, conflicts, and wars. If you should find yourself in such a situation take heart: there are many cases of highly successful people who have worked themselves upward from low entry points.

As an example, we can relate the story of a particular person who spent 2-years seeking employment as a teacher both at universities and schools. He was continually rejected. Eventually, he managed to score a relatively low-level job in another field but he was not promoted in that job. However, he did use the opportunity provided by the job to think laterally about some work he was doing. You may well ask whether, or not, this person ultimately received appropriate recognition and a better job? We leave that for you to judge: His name was Albert Einstein (see Ref. [1], Further Reading).

> Sometimes you may need to accept a lower level position, biding your time until the right opportunity presents itself.

14.4 Writing Successful Job Applications

Having decided where you want to work, the next step is to submit a suitably written application. Be sure to cover the "necessary" and "desirable" attributes of the position. You need to explain why you are ideally suited for the position. There are likely to be many other applicants for the position (sometimes more than a hundred), so you will need planning to make sure that you make the short lists.

> Part of this process is to have a clear and unambiguous Curriculum Vitae (CV) to accompany your application. Make sure it is "to the point" and does not overstate, nor understate, your skills and qualifications.

You should give thought to the special attributes that employees may be looking for in a doctoral graduate. In addition to the specifics of the job requirements outlined in the job advertisement, generic attributes might include the following.

> In universities, the employer is likely to look for general attributes such as
> - Your capacity to do high-quality research.
> - The capacity to attract external research funding.
> - Your ability to work with industry.
> - Your ability to work with others and form teams.
> - Your vision for the future.
> - Your capacity to organize and explain complex ideas.
> - Your ability to deliver a great lecture.
> - Your capacity to motivate students.

> In industry, the employer is likely to look for general attributes such as
> - Your capacity to apply "out-of-the-box" thinking to complex problems.
> - Your capacity to achieve an outcome.
> - Your capacity to represent the company in high-level negotiations or, more broadly, to the public.

- Your ability to organize and plan.
- Your vision for the future.
- Your ability to assemble the resources needed to complete a task.
- Your ability to leverage interactions with universities to multiply the innovative power of your organization.

There is a natural fluctuation, as well as evolution, in what is considered a great application. To some extent this is due to successful techniques being quickly published online. They then become common practice and after a while cliché. Employer's attention is then captured by novel approaches and the cycle begins anew. Therefore, be sure to research current trends, avoid clichés, and prepare well!

14.5 Referees

You will invariably need to have a set of people who are prepared to be contacted to confirm your skills and abilities. Here is where your network of colleagues, especially senior ones, can be a huge benefit. Make sure you ask potential referees politely if they would mind if they are contacted. Give them a "let-out clause", e.g., "I know that you are very busy and hence I will understand if you do not have time to do this for me. Either way please let me know so that I might approach somebody else".

Also, it can be very helpful to send your referees your curriculum vitae so that they have ready access to times and details of your interaction. In some cases, it is also appropriate to send your referees the job advertisement and your written submission. This way they are fully informed about the position that you are applying for and how you meet the necessary and desirable attributes.

Finally, be sure to thank your referees in the event that you do get the job or to tell them that this time you were unsuccessful and ask if they would be prepared to act as a referee again on some future occasion.

14.6 The Job Interview

Once you reach the short list, the organization will almost certainly interview you. Here is your big chance to show you are the person they are looking for. Some associated tips are

14.6 The Job Interview

- Dress neatly.
- Be on time.
- Do your homework so you know about the organization.
- Show passion and enthusiasm without exaggerating.
- Think about possible questions and have clear and unambiguous answers ready.
- Do not over- or under-sell yourself.
- Be direct and firm in your approach.
- Never make negative comments either about this institution (or any other for that matter).
- Do not discuss salary or work conditions unless they ask you (there will be scope for these discussions if they make you an offer).
- Never use sexist or racist language.
- Always be polite.
- Exhibit an air of quiet confidence and inner strength.

Finally, we recall the following overarching advice (Fig. 14.1):

Remember you do not get a second chance to make a good first impression!

Most importantly: be authentic (within the boundaries of the above) and try to stay calm: but not bland, retain your enthusiasm: stress reduces creativity, so practice calming yourself, especially if you have a tendency to become anxious.

Be sure to check the company website and news feeds for any recent major announcements that may come up in the interview. Here is the chance to show that you know what is going on and how your presence will enhance the company's activities.

You should also think about, and prepare for, possible questions that might be asked at the interview. There is actually a set of general questions that are very commonly used. Such questions include

- Tell us about yourself.
- Why do you believe that you are the right person for this position?
- Can you describe a time when you needed to think "outside the box" to solve a problem?
- What do you consider your best and worst characteristics?
- If we were to offer you this position, where would you expect to be in five years time?
- Why did you apply for this particular position?

Fig. 14.1 "Remember you don't get a second chance to make a good first impression" - see Sect. 14.6

Roger Davies (mentioned elsewhere in the book) also gave the following recommendation: "An interviewee will usually be given an opportunity to ask questions. This provides a good opportunity to demonstrate that you have done your homework on the company. A well-considered question can demonstrate an interest in the company's strategic direction, career development opportunities, growth potential etc. However, the candidate must then be prepared for the bounce-back question 'How would you tackle that issue?'"

> Finally, as we have said elsewhere in the book, remember that you should give focused, clear and positive answers. If you want the question clarified, then ask politely. Also, read and think about Sect. 11.3, "Three Minutes is all you have".

14.7 Summary

Part III of the book turns to the early years after you have obtained your doctoral degree. We encourage the reader to read this part as a whole, as later chapters will help you decide whether you want to seek work in the academic, private or government sector.

The opening chapter of this part is concerned with generic aspects of job applications, i.e. aspects that are relevant irrespective of the sector in which the job lies.

One such aspect is whether to "aim high" or "aim low" (Sect. 14.3). In other words, do you seek a position that is a stretch for you and possibly intimidating? Or do you seek a position that you can handle comfortably? Our answer has two components.

First, we note that either aiming too high or too low can lead to poor performance: the former due to stress and the latter due to boredom. So you need to position yourself wisely. Your network of peers and mentors we spoke about in the earlier parts of the book will now be of invaluable help. Ideally, we recommend a position that does challenge you - but not to your "breaking point".

Second, there is a strategic aspect. If you simply cannot land a job at your ideal level then it is worthwhile considering a lower entry "to get a foot in the door" and work yourself up. If you ask around among successful people you might be surprised to see how many of them chose that way at one point or another.

Section 14.4 gives general advice on how to prepare a successful job application. We deal separately with applications in the academic and private sector. Obviously, your application has to be tailored to the specifics of the job. Make sure it is personalized! Also, there is a natural evolution in what is considered the best standard: what was considered a great and innovative style for an application a year ago might feel like an outdated cliché in a few years time. So make sure to research the current trends on the Internet and seek advice.

Job applications usually require you to list referees, i.e., former professors, mentors, or colleagues who are prepared to confirm your skills and other attributes. During the heat of hunting for such referees, Sect. 14.5 reminds you of the associated etiquette.

Finally, Sect. 14.6 turns to the actual job interview. One cannot rank the many pointers by importance, as violating any single one of them can sabotage your entire application. Having paid attention to all of them, the following points bear particular attention: be well informed about the job and the potential employer, know what you have to offer, be authentic, and be passionate without exaggerating. If you tend to be anxious, practice calming yourself since stress reduces your creativity.

Maybe the best summary is given by the quote in Sect. 14.6: "Remember you don't get a second chance to make a good first impression!"

14.8 Further Reading

[1] B. Bryson "A Short History of Nearly Everything" Random House, USA, 2003.

Chapter 15
An Academic Position

15.1 Overview

In this chapter, we give more detail regarding the responsibilities and duties of a young academic. However, not surprisingly, there is substantial overlap with the responsibilities and duties of a person working in industry. We thus strongly encourage you to also read Chap. 16 since much of what is said there also applies to academic positions.

Broadly, an academic is involved in "scholarship". Scholarship can cover many areas including the conduct of your own research, putting other people's ideas into practice and teaching (see also Ref. [2], Further Reading).

The terminology used for different academic positions varies between different countries and continents. Two examples of the scale of academic progression are:

- Postdoc, Lecturer, Senior Lecturer, Associate Professor, Professor
- Postdoc, Assistant Professor, Associate Professor, Full Professor

We will not fixate on the terminology but focus instead on the fact that there is a natural progression. In this chapter, we will focus on the more junior positions (say up to Associate Professor). In Part IV of the book we will focus on more senior positions.

The responsibilities of academic positions cover four core functions: teaching, research, administration and community engagement. The balance of these responsibilities may vary between different positions and the relative weightings can change as one progresses. However, particularly as a junior academic, you are well advised to gain experience in all four aspects so as to develop a balanced portfolio.

15.2 Teaching

As an academic one needs to have a strong commitment to motivating students and helping others learn. Thus, teaching is a core responsibility of the job. "Teaching" is not the same as "knowing". Teaching is about knowledge and skill *transfer* rather than *acquisition*.

15.2.1 Prepare Well

Teaching is not easy. At the tertiary level, the teacher needs to be very familiar with the technology and ideas that underpin the subject. A one hour lecture can require more than eight hours of associated activity in preparation, developing laboratories, marking assignments, etc. We can safely say

> Teaching at the tertiary level is not for sissies.

So what makes a great tertiary teacher? Whole libraries are available on this topic. We suggest that you read some of this literature whilst keeping an open mind. The key thing is that you find what works for you, seek frequent feedback from your students and make sure that they "love" you and your subject. As a starting point, you need to prepare well:

- Prepare well: Know your subject; know how you will present; think about the timing of your delivery.
- Prepare well: Do not overplan your introduction and believe you can improvise the rest (a mistake many a young lecturer has made). You need to prepare your entire lecture, particularly the hard bits!
- Prepare well: Know where your lecture fits into the bigger picture of the entire syllabus.
- Prepare well: Build on where you are coming from and motivate where you are going.
- Prepare well: Assemble the required learning aids and associated technology.
- Prepare well: Guide, do not dominate.
- Prepare well: Pitch to both the left- and right-half brain. Traditional, explicit lecturing, appeals to our analytical left-half brain. Teaching by analogy, story-telling, historical anecdotes and metaphors appeals to the right-half brain. Both can trigger the famous "aha-experience" in a listener. Blending both aspects, without exaggerating, will liven up your presentation and contribute to a fulfilling experience for your audience.

> - Prepare well: Watch both your macro-timing (sticking to the overall time plan) and micro-timing (such as pauses and intonation).

15.2.2 Good and Bad Teaching

> A good teacher needs to want to help others.

> A good teacher needs to know how to acquire and keep the attention of young people.

As Galileo Galilei is reported to have said (as quoted in Ref. [5], Further Reading)

> "You cannot teach a man [woman] anything, you can only help him [her] discover it in himself [herself]".

You may find it useful to reflect on your own teachers from high school and your undergraduate education. Consider which ones you remember fondly and which ones you prefer to forget and why. Oddly enough a single person can be both the best and worst teacher.

Indeed, a young academic friend of the authors when asked to name the best and worst teacher said it was the same person. When we asked how this could possibly be, the person replied, "In one course, the person was full of enthusiasm and joy for the subject. It was extremely interesting and highly motivating. In another course, it was clear that the teacher hated the subject".

> A great teacher shows enthusiasm even in their less favourite subject.

Finally, calibrate your teaching:

> Seek feedback from your students and colleagues.

15.2.3 Focusing the Content

The educator's key role is to "sift the wheat from the chaff", i.e.

> A good teacher needs to be able to distinguish what is important from what is not.

The last point is particularly close to the authors' own personal beliefs. We have the feeling that many tertiary educators, especially in technical fields, over-teach. This has the potential to blunt students' interest and hide the key concepts in a sea of irrelevant detail. We believe that one should focus on basic principles. Metaphorically speaking: to teach the "basis-functions" or "building-blocks" and tools. If students can grasp them at a deep level, it will help them to structure and understand existing knowledge as well as develop new insights, the hallmark of research.

As is probably clear to the reader, the authors of this book completed their own undergraduate education many years ago. Virtually none of the technology specific ideas that were taught at the time lasted more than five years! However, deeper principles (such as electromagnetic theory, modelling, linear algebra, English expression, and physics) remained useful throughout our careers.

> Basic principles outlast short-term relevance.

15.2.4 Communicating with Clarity

> A good teacher selects the core ideas and then communicates them with clarity and enthusiasm.

This raises the question, how does one communicate with clarity and enthusiasm whilst capturing the imagination of young people? Again there are many different techniques. In our careers we came across many brilliant teachers (and quite a few who were less than brilliant).

In this context, we recall the story of the lecturer who asked the class, "Can those in the back row hear me?" A student from the rear of the hall replied, "No, I cannot hear you!" This was immediately followed by a quick reply from the front of the room, "Don't worry, I am happy to swap!" Watch out that this is not happening in your class room.

Another colleague remarked: "Don't worry if you see someone in your audience glancing at their wristwatch. It is when they first glance, then stare in disbelief and then rattle it to see if it is still working - that is when you need to worry!"

On the other hand, some teachers have a wonderful rapport with their students. Three teachers amongst our colleagues who immediately spring to mind as brilliant are Brian Anderson (now at the ANU, Australia), Mario Salgado (UTFSM, Chile) and Arie Feuer (Technion, Israel). Each was tough. Forget the idea of gaining popularity by telling jokes. Each had a unique ability to tell students deep ideas in a clear and simple fashion. Graham recalls a young colleague of his (Katrina Lau) commenting that she could still hear Mario Salgado's voice as he explained Internal Model Control to the class. This may seem unremarkable—save that Mario taught her over 20 years ago!

The important thing is to gain the respect of your students. In this regard, it is helpful to look at what not to do.

How to give a really bad lecture

- Be unprepared.
- Be uninspiring.
- Have poor visual aids.
- Mumble.
- Be disorganized.
- Have no clear message.
- View teaching as a nuisance.

15.2.5 Inspire

Beyond being prepared, having the right content and achieving clarity of exposition, you need to inspire your students. The authors of this book, and many others, can point to one (or sometimes several) teachers, who completely changed their lives. This is the teacher you should aspire to be!

15.2.6 Learn the Lessons of Failure

This section was inspired by comments made by Wendy Mason. Wendy is a Chief Education Officer in Australia. She is an inspiring teacher, a wonderful leader and a gifted mentor. There is nothing she cannot do superbly; well almost nothing. She decided at one point that she would learn how to ride a motorcycle. The associated course, in Australia as in many other parts of the world, is quite difficult and demands a lot of manual dexterity. Wendy found the course quite difficult. She eventually quit - a virtual "failure". Her first ever! How did she respond? After some introspection, she decided that she actually obtained great satisfaction from being the "perfect pillion" on her husband's motorcycle. She said, "That is where I knew I would excel. Also, I loved the feeling of being behind my husband and helping him ride". She went on to say, "I wish I had this experience many years earlier. Learning what it is to 'fail' and how to respond to it would have made me a better teacher".

We recommend that all teachers heed this story. Whether it is failure in your own life or how you deal with failure by others could well define you as a teacher.

> Nobody can excel at everything. Thus helping your students recover from a set-back is an important skill and can define you as a great teacher.

A closely related issue is the recognition that failure can sometimes have dual implications. On hearing that 80% of students had failed a course, one teacher responded, "I'm not surprised, they are all idiots!" Perhaps this is like the wife who called her husband on his mobile and said, "Watch out dear when you drive home tonight, there is some idiot driving on the wrong side of the motorway!" He replied, "What do you mean one idiot? There are thousands of them!" Food for thought?

15.3 Research

Research is a prominent and inescapable part of any academic's job. It is typically 30–40% of the total work load, i.e. an average of one or two full days per week. Around the world, universities are judged on their research performance, with elaborate league tables, selection criteria, comparisons, etc. Thus, you need to get it right. However, if you succeed, then research can be one of the most rewarding activities you can do.

> Good Researchers are enthusiastic and full of life and joy of discovery.

As previously discussed in Sect. 3.4, good researchers ask great questions. Whilst the specific question may change as your career and research evolves, it must always remain clear and unambiguous.

So how does one settle on a good problem/question?

> One must read, think, read, think and then read and think some more.

15.3.1 Good and Bad Research

We believe that

> The characteristics of good research are
> - Elegance.
> - Simplicity.
> - Generalizability.

15.3.2 How to Get Started

So how does one get started?

Our advice is to do something (anything!); get moving. You will get nowhere by worrying.

> The capacity for humans to achieve great things is unlimited provided they try!

Graham often quotes to his students the following piece of Chinese wisdom,

> "Man stand for long time with mouth open before roast duck flies in".

In other words, you need to get going!

Actually, Graham's Chinese doctoral students say they have never heard this saying. So maybe it isn't actually from China. However, it does convey an important message. It beautifully captures the idea that one has to start along a path if one is to have any hope of reaching the end of the path.

> Every long journey starts with a single step.

Graham would also like to recall an encounter that he had with a famous researcher whom he met at a conference in Singapore. He was working on a topic that was far from his background and expertise. Nonetheless, Graham noticed that he methodically attacked the problem. Graham was left with the impression that it was certain that he would come up with fresh and exciting results. Graham was particularly impressed by the way he started. He cannot recall his exact words, but it was something like,

> "Proceed slowly and carefully so as not to miss anything that may subsequently turn out to be of importance".

In the authors' experience, a few approaches that have led to successful research outcomes include:

> - When tackling a problem to which the solution has been blindly accepted by all previous researchers, then approach it from a different perspective.
> - Think laterally.
> - Be willing to question dogma no matter the eminence of the authority stating it.
> - Be willing to recognize that "If something seems to be too good to be true, then it probably is!"

15.3.3 Attracting Research Funding

To do research, one needs external funding. This in turn requires that you apply for grants. Funding buys equipment, pays for postgraduate students and allows you to travel to conferences to present your own work and to hear the latest ideas from others. It also covers your own income and expenses if you do not, as yet, have a position with a fixed salary. In the USA it may be a crucial component in covering your summer salary even if you have a tenured full-time position.

Applying for grants can be enormously time consuming (many hundreds of hours) and frustrating, especially when success rates are often as low as 5–10%.

When applying for grants clarity is extremely important. The following appear, in one form or another, in essentially every grant application proforma.

> Tips for writing a grant application:
>
> 1. Describe the broad problem you wish to study (Aims).
> 2. Explain how it is being done now (Background).
> 3. Explain your "wonderful" new idea (Technical Details).

4. Discuss who will benefit from your ideas and why (Impact).
5. Describe the steps needed to reach your goals (Methodology and Timetable).

You should begin with a one sentence answer to the above questions. Then expand to a brief, but highly focused, response. You should be able to capture the key idea in three minutes. Indeed, we again suggest you review Sect. 11.3 on the value of making a 3-min summary.

One reviewer of an early draft of the book also said that it sometimes helps to be honest about the key risks in successfully completing the research. This demonstrates maturity of thinking.

If you find yourself in such a situation you should then also explain the following:

- How you have taken steps to mitigate the risks.
- Why the suggested project is still the best way forward.
- What will still be learned, if the risks materialize.
- Make it obvious that this is a great proposal to support whatever transpires as the work evolves.

15.3.4 Carrying Out the Research

Once you receive a research grant, you need to carry out the wonderful research that you promised. Remember that this feeds into your reputation which will be the basis of getting your next grant. This has strong connections to the research skills described in Part II. However, an entirely new skill arises when you are a junior academic, namely how to supervise postgraduates and doctoral students so that they do great research with you. To achieve this you need to create an environment in which the students are stretched and stimulated to achieve their best. You also need to be able to recognize when you need to use your past experience to support them in their hour of need.

As you carry out your research, there are a number of questions that naturally arise such as

- Quality versus quantity.
- Should I know "everything about nothing" or "nothing about everything?"
- Should I write one book or ten journal papers?

As is so often true in life, the answer to these questions is a matter of balance. As you develop your own personal balance, remember the importance of collaborations. So, if your answer is, you want to write a book as well as 10 journal papers, your network of collaborators, doctoral students and postdocs becomes crucial. A book will typically summarize a large body of work and is very likely to enhance your reputation and prestige. The obvious caveat is that it needs to be a good book!

You should be warned (from the hard earned experience of the current authors) that writing a book is an enormous amount of work.

The two authors of the current book remember saying to a potential third author of an earlier book ("Control System Design", by Goodwin, Graebe and Salgado, Prentice Hall 2001), "We have a good first draft of the book and expect another 300 h of work should finish it". The truth is that it took, at least, another 6,000 h of work to complete the book! So much for our skill at budgeting work force requirements. The only redeeming factor was that it turned out to be a "good book" as judged by others. Indeed it won the Triennial Best Text Book Award from the International Federation of Automatic Control.

Fortunately, it is true that when one writes one's first book, you are blissfully unaware of how hard it will be. Also when it comes to writing the second and subsequent books, you either forgot how hard the first one was or, erroneously, believe that the second one cannot possibly be as hard as the first.

15.3.5 Time Management

Academics need to be very careful to guard the time they allocate to research. It is of supreme importance, yet it is easily put off in the face of more immediate demands, e.g. those coming from teaching or administration. We advise, firstly,

> Do not put off to tomorrow the research you should have done yesterday.

Secondly, we encourage the reader to also refer to Sects. 23.4 ("Working Smarter Not harder") and 23.5 on approaching time management from an energy management point of view.

15.3.6 Reviewing Papers

In so far as we expect others to review our own grants and papers, we should freely give our time to review the works of others. A rough guideline is to multiply the number of papers in which you are an author by four and divide by the number of senior authors on each paper.

Failure to review papers when asked can have you banned by the journal on the grounds that you are not a "good corporate citizen".

Also one needs to be aware of potential conflicts of interest and either avoid them altogether or, at least, declare them.

> Possible sources of conflict include, but are not restricted, to the following:
> - Knowing very well an author of a paper you are reviewing.
> - Having a disagreement with the author of a paper you are reviewing.
> - Having overlapping research which could lead you to be either too generous or too harsh in your comments.
> - Hoping that, by getting somebody else's work published, you will enhance your own citation count.

When writing a review be as positive and helpful as you possibly can be. In other words

> Write a review that you would be happy to receive yourself.

Show compassion and understanding when writing reviews. The tone matters: avoid emotional charging such as being arrogant, patronizing or ridiculing.

15.3.7 Building Momentum

If your long-term goal is to end up with a major research activity with significant external funding, then you need to be strategic in your thinking. Some important elements include:

- Gather a team.
  - Work cooperatively with like-minded colleagues.
  - Exploit the specific strengths of members of your team.
  - Identify knowledge gaps and work to fill them.
  - Attract top-level doctoral students.
- Choose the right topic.
  - Identify your strengths.
  - Look for a theme where you can make a difference.
- Give maximal exposure to your work.
  - Network as much as you are able with peers, other academics and industrial colleagues.
  - Make frequent public presentations, e.g. at conferences.

- Organize and present tutorial workshops; cover as many sectors as possible including academic, private and government.
- Get involved in conference organizing committees.
- Where appropriate, utilize the media (newspapers, radio and television).
- Have an exciting website.

15.3.8 Multidisciplinary Research

Not surprisingly, some of the very best research occurs at the boundaries of discipline areas. Indeed,

> A new pair of eyes often sees things that are opaque to those who have looked at a problem for many years.

Thus multidisciplinary research can be extremely rewarding and beneficial. Our readers may be interested to know that the first author of this book recently turned much of his research effort into the goal of improving the treatment of Type 1 Diabetes. This topic lies at the boundary between engineering and medicine. This is a jump into unknown territory with associated dangers but also enormous potential.

From Graham's experience over the past two years of doing this work, it seems that the following advice may help others.

People from different areas often view the term "research" in entirely different ways.

> It is easy to be dismissive of others expertise; however, this observation works both ways and unless a good relationship is established early on there is no hope for a brilliant outcome.

There is a "language" barrier that needs to be bridged. Namely, when a person from one side says something it may be interpreted in an entirely different way by people from the other side.

Researchers from both sides need to find common ground. This is a slow process because confrontation generally does not work, although in some circumstances, confrontation is exactly what is needed. Needless to say this requires a fine balance and an understanding of human nature and politics.

15.3 Research

> It is all too easy for the research to "fall between two stools". Thus it is imperative that both sides devote time and effort into becoming familiar with the literature and methodologies of the other field.

It is all too easy for one side to say, "Oh we know all about that already, you are wasting your time looking at it". Be careful; remember that one of the key goals of multidisciplinary research is to "question conventional wisdom".

One of the hardest things to do is to convince the other party to investigate a question to which they believe they already know the answer. This is especially true when the multidisciplinary field is medicine. This is due to several factors including the fact that patients are involved and thus tests that are perceived as being redundant are frowned upon. One possible approach to this problem is to take small steps that highlight what you are trying to do without drastic changes until the unreceptive party comes to their own conclusion that their long held beliefs may indeed be worthy of challenge.

It is helpful to talk through the issues and "put differences on the table". When this has been done, do it again.

> Sometimes it is helpful to think of a Venn Diagram with three sets of "facts". One set contains the "facts" that group number one believe to be absolutely irrefutable. Set two contains the "facts" that group two believe to be absolutely irrefutable. Set three contains the truth. The reader should not be surprised to learn that these sets often have little overlap. Food for thought!

Some areas believe that randomness is a part of the natural world and thus rely upon statistics to justify the unexplainable. On the other hand, engineers and physical scientists tend to believe that everything has a scientific explanation although we may not know the explanation right now. Vasant Natarajan, (see Ref. [4], Further Reading), makes the key point that every scientist must ultimately believe that everything has a rational explanation.

Above all, be open to new ways of thinking, keep an open mind and do not loose your sense of humour.

15.3.9 Making a Quantum Leap Forward

It often happens in research, especially in mature areas that progress becomes rather slow. This is because the research has reached a wall that inhibits further progress. The "wall" may even be preceded by a significant knowledge chasm. In such cases, it is impossible to reach new heights by simply making incremental steps (see Figure

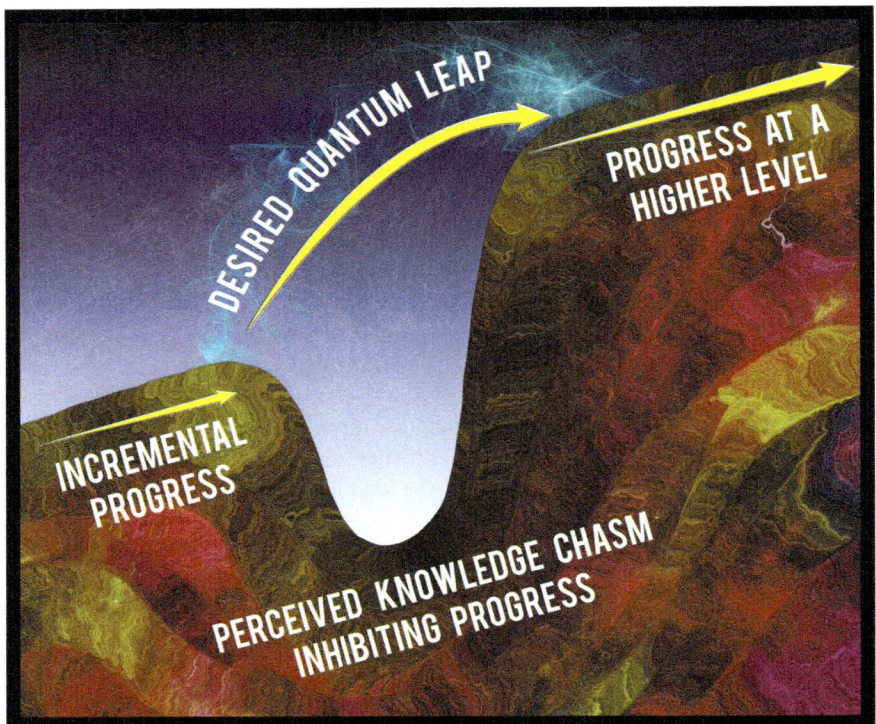

Fig. 15.1 To make a quantum leap forward, you need to look beyond the perceived blockages and jump over the knowledge chasm, see Sect. 15.3.9

Quantum Leap). However, this can represent a great opportunity. To capitalize on the opportunity you need to have a clear vision of what might lie beyond the chasm. Often a "fresh pair of eyes" can see things that others feel are simply impossible. This explains why many big breakthroughs are made by people from outside the particular area. In summary (Fig. 15.1),

> To make a quantum leap forward, you need to look beyond the perceived blockages and jump over the knowledge chasm.

15.4 Publication

One of the key issues related to research is to publish the results. We strongly advise that you reread Chap. 10 on the Art of Publication. The comments made there are equally valid now but with increased intensity and with greater ramifications.

Indeed, your success as a junior academic will undoubtedly be judged by your capacity to write papers and to have others cite them. This is an inevitable consequence of the "academic rat race".

Alas, you have no choice but to join the race. However, we urge you to be strong-willed and to place emphasis on high quality as well as quantity without, of course, succumbing to "self-plagiarism" and other unethical behaviour as warned against in Sects. 12.5–12.8.

> Quality almost always wins out in the long run.

15.5 Administration

Academics are expected to do their fair share of administrative duties. This can be 20% of the work load (say, one day per week). The list of duties can seem endless:

- Assign teachers and teaching assistants.
- Organize rooms.
- Assess potential new students.
- Arrange timetables.
- Assemble examination results.
- Make sure all occupational health and safety requirements are met.
- Address student difficulties, etc.

> The trick in doing administrative work is to be clear about the task and then carry it out efficiently.

Graham recalls the words of wisdom of one of his former Deans (Alan Roberts) who said:

> Administration is either a light meal or a banquet depending on your point of view.

It is important that all academics share the administrative load and not leave their colleagues to do "the dirty work". Thus,

> Be a good corporate citizen and do your share of administrative jobs.

Indeed, seeing administration as being a "dirty job" is going to inhibit you from doing it in an efficient fashion.

15.6 Community Engagement

If you have followed our advice, from Part II, then you hopefully already have a healthy network that you can grow further. Community engagement is simply part of your networking activities. There are many "communities" with which you should network. Some examples are

- Your immediate colleagues
- Colleagues from other departments with overlapping interests
- The broader research community in other universities and countries
- The community in which you live
- The relevant industrial community
- The community defined as professional societies

To underscore the importance of such networking, Graham and his colleagues were recently chosen by a local industry in their home city as their charity for the year (for diabetes research work). A gala ball was held and substantial research funding was forthcoming. Clearly this depended upon extensive prior networking with many groups and individuals.

15.7 Self-Reliance

Being an academic in the modern world is not an easy task. High demands will be placed upon you and you will be expected to deliver on the four areas discussed above, namely; teaching, research, administration and community engagement. You will be expected to deliver at a very high level. Indeed, some universities now have nominal guidelines covering such matters as minimal teaching evaluation scores, number of papers, value of external research income, etc. Also, you will need to be somewhat self-motivated and self-reliant. A senior academic from Princeton in the United States said the following (Fig. 15.2):

15.7 Self-Reliance

Fig. 15.2 "Being an academic in a modern university environment is akin to being self employed". See Sect. 15.7

> "Being an academic in a modern university environment is akin to being self employed. You get a table, chair and desk lamp - the rest is up to you!"

15.8 Working with Industry

The separation between working in academia and industry is becoming much less pronounced than formerly. Indeed modern academics are strongly encouraged to engage with industry. As mentioned elsewhere this can have many benefits including;

- Motivating students
- Being a source of research funding
- Providing exciting and relevant research projects

To work with industry you need to understand their goals and priorities. We thus strongly urge that you read the next chapter.

15.9 Summary

This chapter describes the responsibilities and duties of a young academic. We recommend that you read it together with its companion chapter, Chap. 16, which describes the equivalent responsibilities and duties for a position in the private sector. Comparing and contrasting the duties will give you a good indication of where your inclinations lie.

This chapter is organized along the lines of the four academic core functions: teaching, research, administration and community engagement.

In Sect. 15.2 we look at both good and bad teaching. One can learn equally from either. It is likely that your years of study were more about learning than about teaching. So a first point to realize, for a novice lecturer, is the difference between knowing and teaching: the former is the acquisition of knowledge and skills, the latter is the transfer of them. As you hone your skills as a lecturer, remember that people's minds work differently: something that sounds like a clear explanation to you might well sound like "gibberish" to someone else.

Some of the pointers you will find in Sect. 15.2 include aspects of planning (not just your subject matter, but also micro- and macro-timing, avoiding overplanning the introduction and improvising the rest, etc.) as well as the encouragement to mix analytical lecturing with anecdotal or metaphorical explanations.

The second core academic function, research, is the topic of Sect. 15.3. Of course, you were involved in research during your doctoral studies. So many of the fundamental skills will already be familiar to you: know the literature, think, question, and "see what everybody else has seen but think what nobody else has thought" – see Sect. 15.3.

However, the big difference between research as a doctorate student and research in a full academic position is that you are now responsible for a much wider scope of managing your research, including: getting started on your own (Sect. 15.3.2), attracting funding (Sect. 15.3.3), carrying out the research (Sect. 15.3.4), time management (Sect. 15.3.5), reviewing papers written by others (Sect. 15.3.6), building momentum (Sect. 15.3.7), and capitalizing on the benefits of multi-disciplinary research (Sect. 15.3.8).

Publishing your work is another skill that you have already encountered during your doctoral studies. However, this carries much higher weight in a full academic position (Sect. 15.4). If publishing in peer-reviewed conferences and journals was a recommendation during your studies, it now becomes a necessity as an academic. However, since the advice on writing is largely the same (as given in earlier parts of the book), this section refers you there. Due to the high pressure on publishing as an academic, we re-emphasise the issue of ethics, lest anybody succumb to cutting corners.

Regarding the third core function of academics, administration, our key advice is to be clear and efficient about it (Sect. 15.5). For some young academics, it can be tempting to bemoan and procrastinate when doing these tasks. However, you will find that this will simply cause you to take significantly more time and energy in doing the task. Moreover it will further distract you from your other core tasks.

We also address various forms of engagement: community engagement (Sect. 15.6), being self-reliant (Sect. 15.7) as well as engaging with industry in collaborations (Sect. 15.8).

15.10 Further Reading

[1] K. Bain, "What the best college teachers do" Harvard University Press, 2004.
[2] E.L. Boyer, "Scholarship reconsidered: Priorities of the professoriate" Princeton N.J. 1990, The Carnegie Foundation for the Advancement of Teaching.
[3] G.M. Crawley and E. O'Sullivan, "The grant writer's handbook: how to write a research proposal and succeed" Imperial College Press, U.K., 2015.
[4] V. Natarajan, "What Einstein meant when he said 'god does not play dice...' ", Resonance, July 2008, pp 655-661.
[5] D. Carnegie, "How to win friends and influence people", Simon and Schuster, US, 1936.

Chapter 16
A Position in Industry

16.1 Overview

A position in industry is typically a little different from a position in academia although, unsurprisingly, there is much in common. Hence we will use different headings namely; (a) having and delivering on targets, (b) evolving from administration to entrepreneurship, (c) community engagement, and (d) resources and skills.

16.2 A Doctorate in Industry

When employing a doctorate in industry, the company will typically look for qualities beyond those that a bachelor or master's graduate would normally have. The key issue is that they will expect that a doctoral graduate can be systematically innovative and be able to "think outside the box".

A doctorate in industry will be expected to be flexible, to understand new things, to adapt quickly and to be capable of transferring technology to new areas. Typically, in industry there will be a specification that needs to be met. The researcher will need to find a solution that solves the problem in a practical sense. Frequently, the problem is provided by external circumstances. Thus one usually does not have the freedom to define one's own problem. This is a difference to academia where one can often adjust the problem formulation if the original one turns out to be intractable.

In industry one needs to be focused on applicability and financial viability. The solution needs to "fly" both figuratively and sometimes literally, (e.g., in the aerospace field).

In industry,

- The solution has to work.
- The solution needs to solve the problem.
- It must not infringe upon the Intellectual Property held by other companies.
- It has to be economically viable.
- It has to adhere to all the relevant norms and standards.

Many industrial posts grow into deep specialist roles or technical leadership roles rather than functional management. On the other hand, postdoctorals often make for excellent programme managers. Hence, there is a diversity of roles that one can grow into.

Dr. Roger Davies (mentioned elsewhere in the book) provided the following comments on the expectations held for doctoral graduates in industry.

"Doctorates moving into industry may become involved not only with pure research but also development. I saw a tremendous amount of pure research in the pharmaceutical industry when I worked for Glaxo (now Glaxo Smith Kline). The industry was research driven with massive resources applied to the quest for new drugs for cancer etc. However, more lately (with Nuplex) the focus was more on development of product ranges. This involved meeting with customers to understand what new lines were required in the market and working closely with the customers to modify product formulations to produce the desired end-effects. This required our doctoral chemists to work in multi-disciplinary teams with our own production and marketing staff as well as technical and marketing staff from our customers. Open and honest communication with customers was paramount. Tough deadlines always featured but a nimble, flexible approach generally produced the results and cemented customer loyalty".

16.3 Having and Delivering on Targets

If teaching and research are the core activities of an academic position then delivering on targets and being entrepreneurial are core activities in the private sector. It might seem obvious, but, it is worth remembering that

To deliver on a target, you must have a target in the first place.

Our readers may be surprised to learn how much energy is put into projects and activities with no, or at least ill-defined, targets.

16.3 Having and Delivering on Targets

In addition to having well-defined targets, these targets also need to be properly communicated to the relevant people. If a target is subdivided between several people or teams then one must make sure that the sub-targets are consistent and add up to the whole. In other words, all sub-targets must be aligned with each other.

> Be careful to ensure that the sum of the parts is neither more than, nor less than, the whole—the parts need to add up harmoniously.

This seems straightforward and obvious but it all too often happens that the subdivision of a target is poorly conceived. Also, the available resources must fit the task to be performed.

The skills required to assemble the resources to successfully carry out a task include:

- Having clearly articulated goals (not just a wishy-washy hope) and milestones.
- Assembling the right team.
- Formulating a budget and attracting the finance.
- Assembling the necessary suppliers of goods and services.

On the output side, the skills to deliver both on time and on budget are essential. Moreover, it is particularly when time, budgets and information collide and are at odds with each other and your goals that true management skills are called upon. It is at this time that your doctoral training will prove valuable as it should have taught you how to tackle a complex problem and bring it to a successful conclusion.

There are several popular goal setting systems available at the time of writing this book. Of course, new ones are highly likely to be available by the time this section becomes particularly relevant to you. So be sure to research the current best practice. In the next section we will review one system for goal setting as an illustration.

16.4 S.M.A.R.T. Goals

The S.M.A.R.T. system was mentioned earlier in Sect. 5.3. There are slight variations on how the acronym is interpreted but a common one is the following.

> S.M.A.R.T. stands for
>
> **S**pecific
> **M**easurable
> **A**chievable
> **R**esults-focussed
> **T**ime-bounded

Moving to S.M.A.R.T.E.R., the "E" and "R" stand for

> **E**thical
> **R**ecorded

It is easy to see the advantage of a goal having the above attributes.

The largest difficulty usually arises with the "A" part of S.M.A.R.T., i.e. the goal being "achievable". This is linked to the "R" which is sometimes interpreted to mean that the goal needs to be "realistic". This is, of course, a great property for a goal to have but—particular when research or innovation is involved—it can be difficult to assess.

The S.M.A.R.T. system lends itself particularly well to "contractual goals", in which an agreement is made between an employer and a staff member (or consultant) with financial implications: Usually, the size of financial bonus will depend upon the degree of goal achievement.

We encourage the reader to think about great historical achievements and to think about how "realistic" or "achievable" they were or were not at the time for example, think about Christopher Columbus, the Apollo mission to the moon, or Edison's quest for the light bulb. Conversely, there are also examples of goals that were once considered achievable but are viewed as impossible today for example, think about turning lead into gold or building a perpetual motion machine.

Unfortunately, goals are often associated with uncertainty due to the influence of external factors. One of the ways to deal with these uncertainties is to break the goal down into scenarios as indicated in Sect. 5.3.

Also, if bonuses are attached to achieving certain sub-goals then these need to reflect the fact that something beyond the normal expectation has been achieved. For example, a relatively low bonus might be attached to delivering a project on-budget and on time since it might well be argued that this achievement coincides with the normal expectation. As another example, the manager of a division which achieved a $100 million profit in a particular year might feel poorly treated when awarded no bonus. However, this could be very fair, for example, if the targeted profit was 50% higher.

We end this section with two final pieces of advice. Firstly, do not include incentives for achievements that have no value. For example, if rewards are associated with finishing a project early then maybe there should be some value to this because a new product can go to market. It is very discouraging for staff if they have been encouraged to finish a project early only to learn that the outcome is simply lying around "collecting dust".

Secondly, if certain aspects of performance are measured and rewarded, then it is important to be careful that other aspects are not neglected - sometimes with unintended and detrimental (even disastrous) consequences. For example, you don't want to encourage undertime or underbudget achievement at the cost of quality, safety, legality or morality. The reader is encouraged to think of the impact of disasters and scandals, such as the Space Shuttle disaster or the emission scandals in the automotive industry.

Just as there are formal systems for goal setting such as S.M.A.R.T., there are also formal approaches to achieve an appropriate balance between different categories of goals such as finance, quality, safety, etc. One such system currently in use is the "Balanced Scorecard" (BSC) system. Our readers are encouraged to read about this approach and other related ideas.

16.5 From Administration to Entrepreneurship

One of the key functions of anyone holding a senior position in industry (or, indeed, a university) is to carry out what is broadly termed "management". Libraries of books and papers have been written on this topic. Concepts range from the "common sense" approach to the "scientific or model based" approach. We advocate a middle line including an understanding of both. The article by Ghoshal (see Reference [3], Further Reading) makes interesting reading in this context.

A common misconception is that administrators exclusively administer, managers exclusively manage and leaders exclusively lead. While this separation is sometimes practiced, we emphasize that administering, managing and leading are mindsets that can benefit any position in a company! Leaders sometime need to "merely" administer the delegation of a task and then stay out of it; and administrators sometimes need to show leadership to uncover, say, a looming danger in the bookkeeping.

Below we highlight how you can add value to your position. We set out these suggestions to inspire your own reflections.

16.5.1 Administration

If you have everything you need to complete a task, if you have the target plus the information and all the resources needed, then get going! We consider this to be an administrative task - you administer by ensuring that the available resources and information are appropriately directed towards achieving the goal. We reiterate that this is not the prerogative of an "administrator" alone. Managers, leaders and entrepreneurs all come across tasks where their personal input is largely administrative (for example, administering a delegated directive.)

Administration is usually a large part of positions in the private sector.

> The key thing is to ensure that the way you deal with administrative tasks is (a) efficient and (b) naturally evolves into project management and leadership.

16.5.2 Project Management

"Administration" becomes "project management" when one finds that one doesn't have all the resources or information to carry out the task. Under these circumstances, the job of the project manager is to find ways to *still* achieve the target by coming up with the appropriate measures and to decide what compromises are acceptable.

Whilst you need *skills* when everything proceeds according to plan, special *leadership* is needed to cope with unexpected events. The one thing you can be certain of is that 'the unexpected' will indeed happen and this is where you will have the chance to shine.

Just as being an administrator is not the sole domain of administrators, so is acting like a project manager not the sole prerogative of project managers. A core aspect of project management is to achieve a goal *in spite* of unforeseen obstacles events or lack of resources. This is a mindset that every person, in every role, within a company is well advised to develop.

16.5.3 Leadership

As in the previous sections, we are referring to "leadership" as a mindset. Clearly, it is a skill that formal leaders (CEOs, directors, heads of departments and units, etc.) need to acquire and continuously hone.

Beyond that, however, "showing leadership" is something that can be done at any level and any job in a company. Just as there are some CEOs who merely "administer" their leadership roles, there are also lower ranking positions, in which inspired people show great acts of leadership.

We encourage you to observe and reflect! You learn equally much by observing and emulating the ten top tips for leaders in your role and sector as you do by avoiding the ten top pitfalls that you observe.

Whilst the definitions for leadership *styles* are generally agreed on (democratic, dictatorial, laissez-faire), generic definitions for leadership vary widely. Usually they include components such as

- Giving direction.
- Setting vision.
- Accepting accountability (although this is often sadly absent).
- Charisma.
- Inspiring confidence in stakeholders, including staff, clients and more.

The specifics of leadership can be quite different in different jobs and sectors. Leaders also need to adapt to changing markets, technology and times. However, if

16.5 From Administration to Entrepreneurship

you research, observe and reflect on the top ten tips and top ten pitfalls, you should have a good starting point to inspire you.

16.5.4 Entrepreneurship

Going beyond administration, project management and leadership brings us to entrepreneurship. By entrepreneurship we are not just referring to the traditional definition focusing on new business start-ups. More broadly, we are referring to an "entrepreneurial mindset". This is not the prerogative of any particular sector or level of responsibility. Anybody, in any position, can adopt an entrepreneurial mindset to help solve their problems.

Examples arise in finance, cost cutting, quality improvement, efficiency improvement, developing new products and enhancing existing products. This is where your boss will expect your doctoral training to come to the fore. Note, that this function includes your specific technical knowledge but also many other skills such as your organizational ability, and your capacity to (Fig. 16.1):

> See the whole picture and then step outside of it.

On the output side, you will be judged on the basis of the quality and quantity of the outcomes. At this level of abstraction, there is virtually no difference to what is expected of a researcher in an academic environment.

> Entrepreneurship, or having an entrepreneurial mindset, has much in common with research since it involves many of the skills involved in research, i.e.
> - Asking the right question.
> - Embracing innovation.
> - Managing risk.
> - Managing the inevitable attributes of the "real world", such as uncertainty, lack of information, financial and other constraints.
> - Learning from failure.
> - Having courage.
> - Questioning conventional wisdom.
> - Avoiding status quo.
> - Turning crisis into opportunity.

A person in any position can be entrepreneurial! This includes everyone from the person who cleans the floor to the company director. It's about having an entrepreneurial frame of mind rather than merely administrating a particular job or position.

Fig. 16.1 "See the whole picture and then step outside of it". see Sect. 16.5.4

16.5.5 An Illustration

Say that you are in charge of a project and a crucial sub-task does not perform or complete as expected.

- If you simply run to your boss and report the problem then you are adopting an administrative mindset.
- If you go to your boss and report a set of possible solutions, then you are acting like a "project manager".
- If you engage your boss with a set of possible solutions which have been prioritized then you may be displaying a "leadership mindset". This is especially true if you suggest how each of the possible solutions impacts on time and budget. Also if you explain to your boss how this can be communicated to the various stake holders, then you are definitely showing leadership.
- An entrepreneur does all of the above, and, in addition, derives new ideas and directions which build on the initial failure and experience.

Our readers are encouraged to re-read the example given in Sect. 7.2 on engaging with your supervisor during your doctoral studies. There are obvious and remarkable parallels.

16.6 Community Engagement

This is very similar to the parallel role in academia. We suggest that you re-read the corresponding sections of earlier chapters. In the private sector, community engagement is even more important since the survival of the entire enterprize can hinge upon your networks. It is worthwhile to be broad-minded when you think about who the community includes. In addition to virtually anybody in your own company, it also includes suppliers, customers, government bodies and, even, your competitors. In the latter case, you need to be mindful of anti-trust issues. However, there often exist technical forums where competitors can exchange non-commercial experiences such as best practice, safety issues, innovation and more. We suggest that you be very proactive in seeking positions on such forums as they are extremely valuable for building your network, experience and early career.

16.7 Summary

This chapter has addressed issues related to the functions associated with a doctorate's position in the private sector. It is a companion chapter to the previous chapter, which looked at the academic sector. Whilst acknowledging the overlap, this chapter focuses on the differences.

Compared to an academic position, the topics you will be asked to work on in the private sector are likely to be more specific and more narrowly defined. Usually, the research and development questions that you will be asked to address concern quite specific problems. You will be tasked with finding solutions that are achievable within a certain timeframe, work in practice, are financially viable and adhere to norms, standards and existing intellectual property rights (Sect. 16.2).

A substantial part of the chapter is devoted to setting and delivering on targets, as well as giving a bonus as an incentive.

We have saved the introduction of a formal goal setting system for this chapter, as it is particularly suited to tasks involving small to medium amounts of uncertainty. It is also well suited to so-called "contractual goals", i.e. an agreement between employer and staff with financial implications. These conditions make it more applicable in the private sector than in academic research although financial incentives are often used in some parts of the world in academia.

The system introduced in Sect. 16.4 is known as S.M.A.R.T., an acronym standing for goals that are Specific, Measurable, Achievable, Results-Focussed, and Time-Bound. The section introduced the system and discussed ways of extending it. With all it's practical usefulness, the reader is also challenged to think about historical achievements that sounded less than realistic at some point (e.g., space travel) or others that sounded quite achievable and are today proven impossible (e.g., turning lead into gold or perpetual motion machines).

A second focus of the chapter lies on the difference between administration, project management, leadership and entrepreneurship (Sect. 16.5).

We emphasize that "administrating" is not the prerogative of administrators alone and that "leading" is not the prerogative of leaders alone. We are not talking about the formal job description, but rather about a mindset: sometimes leaders must administer a delegation and then "stay out of it"; and sometimes administrators need to show leadership and innovation. The section explores the differences and encourages you to carry each in your skill set and to gain experience recognizing when and where to apply each mindset.

Community engagement is just as important in the private sector as it is in the academic sector. However, the communities of interest are somewhat different; with clients, suppliers and others now joining the list (Sect. 16.6).

16.8 Further Reading

[1] Y. Olum "Modern Management Theories and Practices" 15th East African Central Banking Course, July 2004.
[2] J. Emblemsuåg "Life Cycle Costing: Using activity based costing and Monte Carlo methods to manage future costs and risks" Wiley, New York, 2003.
[3] S. Ghoshal "Bad management theories are destroying good management practices" Academy of Management Learning and Education, 4(1):75–91, 2005.

Chapter 17
Moving Freely Between Academia and Industry

17.1 Overview

There are many parallels between a job in academia and a job in industry. The core activities are closer than may be apparent at first glance. To illustrate, we discuss the careers of two individuals; one who made the transition from university to industry and then back to university; the other who held a full-time industrial position but also an adjunct professorship in a university.

17.2 Moving Freely from Academia to Industry

There is much in common in today's world between holding an academic position and working in industry. We know many highly successful people who have been able to move freely between the sectors.

When going from academia to industry a person with a doctorate may need to "eat some humble pie". Dr. Torbjörn Wigren (Senior Research Engineer with Ericsson AB) made the following comment: "Going from Academia to a high technology industry, such as a telecommunication company, can be tough. Even a full Professor of Telecommunications is likely to know less than a junior engineer about the way systems really operate in practice. Professors may be surprised to realize they are the ones in the group who actually know the least".

If you do transition from academia to industry then you will have accumulated particular skills and perspectives in your previous position. These can be an advantage in the new sector but they can also be a "blind spot".

An academic moving into the private sector, may bring important skills such as:

- Ability and courage to question prevailing wisdom.
- The confidence to challenge traditional external suppliers' knowledge, equipment, etc.
- Experience in dealing with complexity.
- Perseverance and attention to detail.

The blind spots that academics going into positions in industry may need to watch out for include:

- Being too scholarly or purist.
- Focusing too much on technical purity.
- Underestimating the importance of leadership and management skills.
- Underestimating company dynamics and tactics.

17.3 Moving Freely from Industry to Academia

Going the other way, from industry to academia, can also be tough unless the person has appreciated and developed the academic context of their work and translated the work into publications.

Some of the important skills that may be carried from industry to academia include:

- Leadership and management experience.
- Strategic thinking.
- Delivering on promises in a timely and cost effective fashion.
- Seeing the relevance and context of research in practice.

The blind spots may include:

- Underestimating the rigorous requirements of academic research.
- The high priority placed on publications.
- Underestimate the importance of technical purity and elegance.

- Overemphasis on time and budget constraints at the expense of achieving a breakthrough.
- Underestimate the importance of seeing your work in the context of existing literature.
- Misunderstanding the importance attached to "academic freedom" within the academic sector.

17.4 A Case Study of a Person Who Successfully Made the Transition in both Directions

The following commentary was provided by Prof. Adrian Wills. Adrian made the transition from academia to industry and back. There are many important aspects discussed in this commentary:

"Moving from academia to industry, and back again, was both painful and rewarding. After more than ten years working as a research academic, I had an almost uncontrollable urge to apply some of my knowledge within a more commercial setting. I believed that the best way to quench this desire was to obtain a job within industry.

At the same time, my academic career was beginning to grow rapidly and I had been offered an ongoing position at a university; which is often considered the "holy grail" of academic positions. Although grateful for the academic offer, I turned it down in preference for a job within a local engineering company. To some, this decision may seem a little odd. I think you are right. Yet, I trusted my gut and went regardless.

I quickly discovered that industry is primarily driven by the bottom line, but not exclusively. This means meeting deadlines, and meeting them on, or under, budget. When combined, these two pressures often result in engineering solutions that are predictable, robust, cost effective, and often very clever.

At the same time, these pressures rarely allow for further discovery and a fresh approach to the problem. This is acceptable when there is no need to research a new approach or to develop a new product. On the other hand, creating or breaking into new markets almost certainly requires some investment into research and development activities.

Fortunately, the company I was working with had a great outlook towards R&D and invested significant funds towards growing in new strategic directions. Over the course of three years, I was allowed to become an integral player in this R&D activity and ended up the Director of R&D.

Out of curiosity I asked the company Directors why they chose me, to which they stated that my training as a research academic was one of the primary reasons. Fantastic, I thought, the last ten years weren't for naught!

These R&D activities have proven to be very lucrative and the R&D team was (and is) the most rapidly growing section of the company. Once again, just while things were heating up, I was presented with an opportunity to leave industry and return to an ongoing academic position. This was a tough decision. As the R&D Director I was really enjoying the technical challenges and leading the team to some successful outcomes. Life was good. Why would I change it now? My logic for leaving industry and returning to academia was primarily based on the following ideas:

- All of the R&D activities that I was involved in could be happily accommodated within a university environment;
- As an academic, having close connections with industry means an almost endless supply of challenging and relevant problems;
- I strongly believe that universities are a great place to conduct industry supported R&D because of the possible leverage of funds available through grants;
- I believed that my industrial experience may be beneficial to current and future students.

I have now returned to the University as an academic, for which I am grateful. I am maintaining strong connections with industry, being involved in several ongoing and some new R&D directions. Already, we are seeing the benefit of this new working relationship with a number of doctorate students having direct access to industry problems with accompanying financial support. To me, this is a fantastic way for the commercial sector to enjoy the potential benefits of quality research, without paying a premium for it. I do believe that this is a good risk mitigation strategy.

My current involvement with industry is necessarily changing from research and development/implementation to that of research only. Rather than deliver the final product, I believe that it is my job to deliver the fundamental knowledge and direction. After all, industry does not want to rely on academics to maintain and develop products and solutions for their customers, this is best served by the supplier. Therefore, my involvement has changed and, depending on the project, this involvement may vary from consulting work to directing doctorate students, directing industry staff, and combinations thereof.

I will conclude with some reflections: I am strongly of the opinion that I am privileged. I was privileged to be offered the chance to do a doctorate, especially with such great mentors. This opened many doors, both academically and commercially and I am truly grateful. I was privileged to have postdoctoral work within the university for some ten years, in which time I travelled across the globe and worked with some great people. I was privileged to be offered a tenured job in a university, only to turn it down and accept another some years later. I was privileged to be offered a job in local industry, and I now have some excellent opportunities to continue working with them. My challenge, as I see it, is to ensure that I provide others with this opportunity and give back".

If this story inspires you, then sharing it in this book has become part of Adrian's wish to "give others this opportunity and to give back".

17.5 Holding a Joint Appointment Between Academia and Industry

One way to facilitate moving freely between industry and academia is to hold a joint appointment, e.g. a full-time industrial position and an adjunct or conjoint position at a university. This arrangement is very common in medicine but less so in engineering and the physical sciences. Nonetheless, we see advantages in this form of appointment.

> Some advantages of joint academia/industrial appointments include:
> - Providing a context for academic research.
> - Access to a university "think tank".
> - Providing opportunities to engage students in industrially relevant research.
> - Opening the door to follow up publication opportunities that may not exist in industry.
> - Having access to good students for future employment opportunities in the company.
> - Producing long lasting results rather than a short-term solution to a problem.

We know several people who benefited greatly from such an arrangement. For example Dr. Torbjörn Wigren (mentioned earlier) in this chapter holds a full-time industrial position in industry and an adjunct Professorship at Uppsala University.

Torbjörn began work for Bofors in Sweden. He spent three years there. Meanwhile he took doctoral qualifying courses completing 2/3 of them on a part-time basis. He then took up a full-time doctoral position at KTH in Stockholm where he worked with Torsten Bohlin, completing his doctorate in a further 18 months of study. He then joined Ericsson Radio Systems where he mainly worked on audio signal processing. He then moved through numerous positions at Ericsson AB including activities connected with 3G, 4G and, most recently, 5G mobile telecommunications.

He holds many patents (more than 150) and was named as the Ericsson AB inventor of the year in 2007. He also has many hundreds of conference and international journal papers.

17.6 Summary

This chapter has discussed the potential of moving freely between academia and industry. Moving from academia to industry is possible but can involve a reality check, e.g. even a full professor from academia may find they know little about certain details involved in the industrial position (Sect. 17.2). Moving from industry to academia is also possible but it is desirable that the person involved have prior experience with academic values and expectations (Sect. 17.3). Section 17.4 provides a case study of a person who successfully made the transition between academia and industry and vice versa. This section contains many positive messages including how the skills acquired in one sector can have flow-on benefits if applied appropriately in a different sector. Finally, Sect. 17.5 presents the case for holding a joint appointment between academia and industry.

Chapter 18
The Cycle of Success

18.1 Overview

Once you have chosen where you want to work and have landed your first job, then you can turn to thinking about how to *grow* your career. A useful way to think about this phase of your professional life is as a cycle of the three elements shown in Fig. 18.1: position plus reputation, skills plus resources and leadership.

Your *position* not only covers your formal job description but also the full spectrum of your professional roles. These link very tightly to your *reputation*.

Your *skills* include the technical skills gained from your doctorate but also the associated skills of presenting your ideas and applying for funding. These link tightly to the *resources* available to you including your team, equipment, budget, etc.

Your *leadership* includes your capacity to draw others to work with you. It also includes your ability to mentor and develop others as well as yourself. Finally, a core aspect is your capacity to envisage and achieve milestones.

If you manage to link all three elements shown in Fig. 18.1 in an upward evolving cycle, then they will feed into each other to accelerate your career. As a new doctoral graduate, all three aspects will probably be in a relatively immature state. Your future development then depends on getting the cycle of success to accelerate upwards. As your reputation grows it will allow you to hone your skills and resources. These in turn, will allow you to develop your leadership skills which will then enhance your position and reputation. With some adaptation, these concepts hold equally well for the university, private and government sectors.

There exist entire libraries written on each of the above three aspects. Our goal in this chapter is to give you specific pointers that will be of particular relevance to building on your doctorate as a starting point.

Before we embark on this journey, a word of caution. Just as the above cycle can be put into an upwardly accelerating mode, it can also easily fall into a downwardly decelerating mode. The latter can occur in many ways, e.g. you might irreparably

Fig. 18.1 The "Cycle of success"

damage your reputation by unethical behaviour of some sort. Unfortunately the cycle is virtually impossible to keep in equilibrium. Hence,

> You will either slowly and surely wind your cycle of success upwards or downwards. So choose wisely!

We treat each of the elements shown in Fig. 18.1 in the sequel.

18.2 Position Plus Reputation

The first element of the circle of success is the combination of position plus reputation.

How you choose to engage in your position is closely linked to your reputation. It is worthwhile to think of various aspects of your reputation such as dependability, versatility, breadth, depth, honesty and being a team player. Some of these aspects can be pursued directly whereas others are a consequence of your actions. For example, you can work on your breadth of knowledge, whereas respect is something your earn as a consequence of behaviour.

18.2 Position Plus Reputation

Fig. 18.2 "Reputation is like a crystal vase, it is beautiful to have but easily broken". - see Sect. 18.2

> You only have one reputation so guard it carefully.

Also remember that (Fig. 18.2),

> Reputation is like a crystal vase, it is beautiful to have but easily broken.

Whilst it is difficult to give specific advice on reputation, we mention it here because of its pivotal importance in getting your cycle of success turning in an upward direction. Your reputation will determine whether or not you will be considered as appropriate for future opportunities and whether or not you will be successful in your applications for positions and resources.

18.3 Skills Plus Resources

The second element of the cycle of success is that of skills plus resources.

In any position you need the skill to both carry out the job and to attract the necessary resources. Virtually everybody in their career is aware of the skills necessary to do the job but are less aware of the skills necessary to attract the necessary resources.

In the following subsections, we discuss different types of presentations (or "pitches") that can be used to acquire additional resources.

18.3.1 Making a Spontaneous Pitch

An important aspect of attracting resources to your project is your capacity to capture complex and difficult ideas in a succinct fashion so that they are easily understood and appreciated. You never know when you will be called upon to present a case for something you want to achieve. We loosely call this, making a "spontaneous pitch".

Previously, we have talked about the importance of the three-minute presentation. The "spontaneous pitch" builds on this idea and extends it.

We relate the following personal experience of a "spontaneous pitch".

In one of his first positions as a manager in the private sector, Stefan was the Head of the Department for "Advanced Control and Optimization" in an oil refinery. About two weeks into the job, there was an unannounced visit by three gentlemen to his office. These were introduced as the plant director, the business unit director and the chief executive officer (CEO)—comprising the three layers of organizational hierarchy to which Stefan reported.

The plant director explained that they were on a personal tour. He then turned to the CEO with the words: "I have already had three different managers in this department during my tenure as director here, and none have ever been able to explain what they do. However, we keep giving them some small funding because their work sounds important. I doubt that Dr Graebe here will be able to explain their activities, especially as he has only just joined us".

An awkward silence followed, with the senior managers looking taken aback. They were clearly a little shocked although not necessarily at the expense of Stefan since he was only a recent recruit.

The opportunity almost passed in mutual paralysis—when Stefan grasped it with a spontaneous pitch.

"Well, gentlemen", he started, "this department does the following: it earns the business money. My colleagues and I carefully develop formulas that we encode in your computers. They continuously adjust hundreds of valves, temperatures, flows and pressures".

Stefan pointed at the complex array of distillation columns and pipelines outside of the office window.

"Since I started talking just now, about 35,000 such adjustments have taken place. All of them carefully with three goals in mind: to reduce energy consumption - which reduces our huge energy bill; to improve product quality - which reduces our bill for wastage; and to increase product yield - which improves earnings".

Again a silence followed. However, this one was not the least bit awkward anymore. The silence was broken by the plant director smiling and saying: "That is the explanation I have always been waiting for!" The CEO nodded thoughtfully, then he said: "And now I think we should not keep you any longer from earning money for the refinery! But", he added, "I do expect a proposal on my desk regarding what additional resources you would need in order to double the success of your department!"

For the next financial year, Stefan submitted a 1 million Euro budget proposal to "multiply the contribution of advanced control to the company's refining profitability". When the investment was earned back within a year it was a major boost to Stefan's subsequent management career. And the three-minute pitch turned out to be a pivotal moment - a moment he could have easily missed.

We have titled this section "spontaneous pitch". To be precise, it is the moment of delivery that is spontaneous and unprepared. An opportunity arrives unannounced and you pitch.

The content, however, is well prepared. In Stefan's case the pitch came form an old habit of his to always ask himself (no matter what field, what project, what paper or client he is working with) how could I summarize this in three minutes?

The main purpose is actually to get to the essence of the matter in preparation for a positive spontaneous meeting. Even if the pitch is never pitched — the process of developing the pitch serves an important purpose in its own right.

Of course, if a spontaneous opportunity does arrive, such as three senior managers turning up at your office—then it sure comes in handy to have a pitch "in your back pocket", ready to go.

18.3.2 Making a Prepared Pitch

Next we consider the alternative "prepared pitch".

Imagine that you are seeking funds from an outside organization and you have just 50 min to convince the decision-maker. Then you need to be exceptionally well prepared including being dressed appropriately, being rested and on time. Recall,

> You do not get a second chance to make a good first impression.

It is also often helpful to have written briefing notes which you leave with the decision-maker.

> Briefing notes can be structured with headings such as
> - Background.
> - The opportunity.
> - Related work of other groups.
> - Our strategic advantage.
> - What we can deliver.
> - Benefits.
> - What I am seeking from you today.

If you are making a formal presentation, then you should structure your presentation in three sections:

In the first three minutes you would give the broad overview and briefly outline the structure of the remainder of the presentation. Then you would go through your key points, say as listed above. Finally in the last three minutes you would give a clear and unambiguous summary of what you have said, what you plan to deliver and the support you are seeking to do so.

If you put yourself in the shoes of a busy decision-maker who is naturally wary of tedious and rambling requests, would you not feel well taken care of if a presentation of the above general format were given to you? Also, note that the same technique works well for both external and internal decision-makers.

Sometimes a team of your colleagues will be involved in the interview process. In this case, it is wise to form a clear view of the "what", "how" and "who" of answering the questions. Specifically

> - Plan *what* you want to say.
> - Be clear on *how* you want to say it.
> - Be aware of *who* will say it.

Indeed, it is often a useful strategy, to have a senior person from your team field all questions. They can give a brief (in say, 30 s) broad response and then pass the question onto another member of your team to give a more detailed response (in say, 2 min). The advantage of this strategy is that it gives the final responder a few valuable seconds in which to frame the answer, i.e. to think about the "*what*" and "*how*". Also, the initial responder can steer the "*who*" without running the risk of somebody else jumping in and giving a poor answer.

18.3.3 *Attracting the Right People*

A key resource that is needed on every project is that of people.

The ambition should always be to aim to attract, and work with the absolute best people. This can be a scary prospect especially at an early stage in one's career. One thing you may worry about is the possibility of ending up in their shadow. However, this thought should be resisted since it is only a team made from the very best people that will be the "winning team".

The first step in attracting good people is to know in which sense you want them to be good. An excellent way to do this is to look at the task at hand, consider your own skills and think how best to complement those skills to carry out the task. You might require different skills for different tasks. For example, one project may require particular technical skills whereas another task may require organizational or presentation skills.

> Broad skills, which go beyond those necessary to carry out a specific task include:
>
> - Being able to work in a team.
> - Being a good communicator, both oral and verbal.
> - Having a strong work ethic.
> - Being goal orientated.
> - Being ethical and honest.
> - Having specific language skills.

One of Graham's former doctoral students (Turker Ozkocak), who now works for Shell in the United States, made the following observation:

"Later in my career, I regretted not having studied Spanish much earlier while I was surrounded by Spanish speaking fellow doctoral students. While talking to a South American operator, this is the only way. Earlier is always better. I wish I had started taking language classes at the university".

> Remember that broad skills may be every bit as important as specific technical skills.

18.4 Leadership

Leadership is the third element of the cycle of success. Naively one may feel that leadership is a simple matter. However, it is anything but simple. You are encouraged to reread Sect. 16.5 which contrasts administration, project management, leadership and entrepreneurship. Additional features of particular relevance to the cycle of success are discussed below.

18.4.1 Team Development

Having attracted the best people, the next task for the leader is that of creating an environment in which the team can be nurtured. This includes respecting the diversity of skills that you have attracted without being laissez faire about the need for leadership.

> Amongst other things, this means creating a healthy environment in which people are not afraid to make mistakes and in which people can grow and mature.

18.4.2 Dealing with Mistakes

Mistakes are not necessarily a bad thing provided they are part of an essential learning process and provided, of course, that the consequences are not catastrophic. Indeed, one should encourage junior staff to not be afraid to make the occasional mistake. Graham would like to add a small personal anecdote about making a mistake.

One of the forms of recognition he was fortunate to receive during his academic career was to be elected as a Fellow of the Royal Society, London allowing him to put the initials FRS after his name (see also Sect. 9.7).

When one joins the Royal Society one is asked to sign a very famous book. Actually it is exactly the same book signed by earlier members of the Royal Society including Isaac Newton, Christopher Wren, Charles Darwin and many others. Graham was deeply honoured and humbled by his election to this society. On the day of induction to the society, one actually signs the famous book with a quill pen (for those who do not know, this is a pen made from a feather that you dip in ink). Of course you need to practice and the people at the Royal Society say, "Whatever you do don't make a blot with the ink!" Easier said than done?

Fortunately Graham survived the signing without making a blot but one of the researchers who signed after him was not so lucky. He made a rather large blot. If he could have died he would have willingly done so! Anyway, the helpers rushed to the stage and said to this poor chap, "Look, it is ok. Actually, it is the blots that make this book real!"

Our readers may like to reflect on the importance of taking bold steps in life even if there might be the remote chance of making a blot. Our advice is as follows (Fig. 18.3):

18.4 Leadership

Fig. 18.3 "Be courageous and don't be afraid to make the occasional mistake. Remember it is the occasional blot that makes it real!" - see Sect. 18.4.2

> Be courageous and do not be afraid to make the occasional mistake. Remember it is the occasional blot that makes it real!

Deciding which mistakes are either (i) incidental, (ii) helpful in the process of growing a person or (iii) are worthy of criticism, is again a key component of your role as a leader and mentor.

18.4.3 Mentoring

Just as it is useful for your team to be mentored by you, it is also important that you seek mentorship from an appropriate senior colleague. You may have to play an active role in seeking this mentorship since you need to remember that senior people are typically very busy and may feel that you fall naturally under somebody else's umbrella. However, having a supportive and knowledgeable senior mentor, can not only be very important for your own growth, both technically and personally, but it can also be extremely satisfying and enjoyable. In the authors' experience, it is amazing what mentorship can do to enhance one's self esteem and capacity to succeed.

As a specific example, Graham, recently decided to take on the role of mentoring a more junior colleague working in the area of Power Electronics. Graham did not have specific expertise in this area but felt his broader view could be helpful. The mentoring helped the colleague secure a very competitive research grant. However, the interaction was even more beneficial for Graham, the mentor. It introduced him to a new and exciting area. It was also hugely rewarding to be able to discuss novel ideas with a bright young mind. The message is:

> Mentoring should bring positive rewards in both directions, i.e. to the one being mentored and to the one doing the mentoring. If it does not, rethink. No matter whether you are the mentor or the mentee.

18.4.4 Friends and Enemies

We have frequently alluded to the importance of both your team and your network. These people will typically be perceived by you as your "friends". However, it is almost inevitable that you will encounter people who are not quite so positive about you or what you do. You should resist the temptation to view such people as your "enemies". Instead you should see this aspect of life as a great opportunity to demonstrate your leadership skills by turning the situation into one of growth.

18.4.5 Dealing with Rejection

One form of antagonistic interaction comes in the form of a rejection. These are never fun but everybody has experienced one at some point in their career.

However, a rejection need not be the "end of the world".

We wonder how many of our readers will know who we are talking about in the following anecdote.

At primary school this person was regarded as an average student. Later at high school, he received a C in Music (his only failure) and was told he had no aptitude for singing. One of his first jobs was as a truck driver. He tried out as a singer for a band but was told he should stick to driving a truck. Can you guess who it was?

18.4 Leadership

This person went on to have the greatest number of songs charting in Billboard's top 40 and top 100. Did you guess? It was Elvis Presley!

> The moral of this story is that you should take a rejection as an opportunity for growth. Indeed, how one interprets a rejection and what one learns from it, are measures of your true leadership capacity.

18.4.6 Lead by Example Not by Brute Force

> People respond well to leaders who act according to the principles they expect of others, e.g.
>
> - Respecting the rights of others.
> - Honesty.
> - Integrity.
> - Hard work.
> - Straightforward interactions.
> - No hidden agendas.
> - No favourites amongst staff.

18.5 Summary

The final chapter in this part of the book introduces what we call your "cycle of success".

It consists of the three elements "position plus reputation", "skills plus resources" and "leadership". As a young doctoral graduate, all three elements are likely to be relatively undeveloped. The goal is to start evolving the cycle so that each component feeds into the other components to create career growth: skills plus resources allow you to produce results that grow your position plus reputation, which is the basis for more skills and resources. With leadership you manage the cycle and give it direction, thereby making leadership an integral element of the cycle (Sect. 18.1).

Section 18.2 examines the role of position plus reputation in your cycle of success. In the academic sector, a Professor holds a more senior position than a Lecturer. In a company, the Head of Department holds a more senior position than a staff member. In either case, a person holding a more senior position tends to be better placed to garner resources (financial, staff, equipment, mandates, etc.). However, it is your reputation that plays a key role in getting you to that senior position in the first place. Here we can already see how position, reputation and resources feed into each other.

As far as reputation is concerned, we encourage you to think beyond the common "good" or "bad" reputation. Other important aspects of reputation include being dependable, innovative, and versatile, having the right mix of breadth and depth, and being able to combine theoretical and practical skills.

We mention these qualities without fundamental judgement: the world needs all of them. It is more a question of how you would like to position yourself. Some of the qualities of reputation can then be actively pursued (such as breadth of knowledge), others have to be earned (such as respect).

Turning to skills plus resources, Sect. 18.3 highlights the importance of your communication skills to "pitch" to decision-makers in order to attract resources. It is always advisable to have a three-minute pitch in your "back pocket"; you never know who you may meet and what opportunities might come your way. The importance of being prepared for scheduled and longer presentations goes without saying.

Attracting the right people (Sect. 18.3.3) is squarely on the borderline of skill and leadership. In the early years of your career, it is easy to underestimate this skill. Dealing with talent can feel threatening. This may tempt you to shy away from making challenging appointments. Ask any senior professional about the importance of attracting talent. They will always confirm the importance of choosing the right talent for long-term success.

The third element of the cycle of success is that of leadership (Sect. 18.4). Ultimately, all elements feed into each other. The specifics and magnitude, however, evolve as your career matures into more senior years. This will be the topic of the final part of the book.

Summary of Part III

This part of the book has highlighted how your position, reputation, ability to attract resources, capacity to deliver on promises and your leadership skills all feed into each other to launch your early career. With small variations these guidelines are true for university, government and private sectors. We encourage our readers to take a break from being submerged in their work and rise above their daily routine to reflect on these topics. This will then quickly become second nature and your cycle of success will accelerate upwards.

Part IV
Using Your Doctorate: The Later Years

Overview of Part IV
In this part of the book, we move onto times when you approach a more senior position. You will now command a higher level of authority and larger resources. These will influence how you put into practice the lessons and experiences you have already gained. Now is the time to think about inspiring the new set of graduates, young professors in your university or junior managers in your company to achieve their own goals and to ultimately become the successful new generation of leaders advancing research, theory and application, knowledge and innovation.

Chapter 19
The Cycle of Success: The Later Years

19.1 Overview

The principles of the cycle of success described in Chap. 18 remain valid throughout your career. However, both the content and intensity amplify as your career matures. It is therefore appropriate that we review the cycle in this chapter and give further emphasis to issues that you may confront in a more senior position.[1]

19.2 Position Plus Reputation

As your career advances, you might find that, instead of managing yourself or a small team, you will advance to managing a large team and ultimately to become a manager of managers. For example, in a university environment, you might advance from managing a small research group to becoming Head of a Department. Eventually you could become Dean of a Faculty or even a Vice Chancellor or Principle of a University. In industry, your responsibilities could advance from managing a small department to managing a division comprising many departments. You could ultimately advance to becoming an executive of one or several companies.

In this context, your reputation from early years can be crucial to your survival. It is well known that many senior people suffer a catastrophic failure due to earlier reputation damage. Indeed, on a daily basis the media publishes details of senior people who have lost their position, (and often much more), due to various indiscretions. Sometimes it is indiscretions from early career years that catch up with them in their senior years when they become more public figures. On the other hand, having a trustworthy and respected reputation will protect you if false or ill-intended accusations arise.

---

[1] If you are not yet in a senior position, then this chapter will provide valuable "stretch information" showing you how your career may evolve. The time to prepare for this evolution is now!.

In a senior position, it is not only your reputation that is at stake, but you are also accountable for the reputation of the entire team that falls under your responsibility. You must therefore have mechanisms in place which ensure that ethical behaviour is part of the ethos of the entire organization. A damaged reputation, even at a relatively low level, can have repercussions right up the management chain.

Also, if an incident does occur, be very mindful of how you manage the ensuing fallout. There are numerous examples of people in senior positions who are not guilty of an indiscretion themselves but who suffer from the way they handle an indiscretion by another. We advise it is always best to be "up-front" and "honest". This will "nip things in the bud" before matters get out of hand. Attempts at damage control by a cover-up typically result in a "tsunami" down the track. The reader may like to think of their own examples including examples which reach all the way to certain Presidents of the United States of America. No-one is immune!

19.3 Resources Plus Skills

As in your early career years, your more senior years still require the necessary skills to both obtain resources and to deliver on the associated promises. However, as you command ever-greater resources, the stakes become ever higher and new skills become necessary. Thus, the skill to pitch your ideas to obtain funding becomes more difficult as you advance up the chain of authority. You will now be more likely to be involved in much more complex issues. You will need to summarize these in a fashion that is readily appreciated by senior decision-makers.

If you are in a university then you may progress from applying for grants that cover specific research projects to large centre proposals that cover multiple inter-related sub-projects. The same general principles apply as were discussed in Sect. 15.3.3, but now with a much broader focus. We suggest you begin with the five questions set out in Sect. 15.3.3. Then you may wish to expand your planning to include other topics.

> Important issues that you may wish to consider when applying for a large grant include
>
> - What is the focus and scope of the proposed work?
> - Why is this particular grouping of projects important?
> - Who will be included in the team and why?
> - Will the team be spread across multiple disciplines or universities?
> - What are the scientific credentials of the proposed team members?
> - Why is the proposed team the best on the planet to do the work?
> - Why does the proposed director have the necessary scientific and leadership skills to lead the centre?

19.3 Resources Plus Skills

- What external groups of collaborators will be included?
- Will your team need to acquire new skills, e.g. by making new appointments?
- What management structure will operate?
- How will the individual research projects be structured?
- Who will lead the individual projects?
- What is the budget?
- How will the funds be allocated?
- What specific outcomes will be delivered?
- What internal Key Performance Indicators will you establish?
- What new resources (people, equipment, space) are needed?
- How will external input be obtained, e.g. will there be an external advisory board?
- Where are the strongest external competitors located?
- Why does the proposed centre represent incredible value to the funding agency?

The above questions were formulated for the university environment. However, they apply, with only small adaptations, to the industrial environment.

19.4 Delivering on Promises

With regards to delivering on promises, always remember that you are ultimately accountable for the outcomes. However, now you may not be directly involved in carrying out the work. Thus, you need to manage others to achieve the promised results. Eventually you will manage other managers who will direct the resources.

19.5 Efficient Conduct of Meetings

Inevitably, you will need to meet with others to gather information and to coordinate resources. In this context, you should be aware of the inefficiencies that can arise from meeting with your senior management team. For example, if you gather eight people for a one hour meeting, then every hour translates into a person day. If you gather 40 people, then every hour translates into a person week. Hence you need to be aware of the resources you are tying up and act accordingly.

Careful planning of meetings is absolutely necessary. Always have an agenda and plan the issues that require decisions to be made.

Be mindful that long meetings typically achieve less than short meetings. In our experience, 50 min is a better defined time period than an hour. Also, adapting the three-minute pitch technique from Part III can prove very useful. Rather than long meetings, it is almost always preferable to package the items in appropriately sized bundles that can be resolved in 50 min followed by relevant action items. If you have ever "died" of boredom in a meeting then you will know you should not impose this fate on your colleagues!

Books have been written on this topic, but it invariably comes down to careful preparation, focus and commitment.

19.6 Leadership

We have argued before that leadership underpins your early career. Leadership becomes even more important as you advance to higher levels. In a large team you will have to deal with unexpected adversities at many levels. Thus developing the right team culture becomes critical. Moreover the driving force behind that culture always rests with the "top dog". In a large team, you will probably no longer lead through your explicit technical expertise but by setting the overarching standards that govern excellence and the achievement of goals.

We encourage the reader to review the earlier discussion concerning leadership in Sects. 16.5 and 18.4. To those earlier comments we add the following.

19.6.1 Always Appoint the Very Best People

If you want to have a successful team then it must contain the right people. Imagine choosing a soccer team to compete in the World Cup but not having the absolute best goalie that you can lay your hands on in the team.

You should always aim to choose people who have great potential. If you are not doing that then there are only two possibilities, namely (i) you have an inflated view of your own abilities or (ii) you are unprepared to be challenged by an "up-start".

Also, keep in mind to balance different personality types in your team. If you are a great visionary, make sure you also have implementers on your team. If you are more of a risk-taker, it might be prudent to have a cautious team member as well.

Inexperienced leaders tend to brush away contributions of personality types other than their own; they feel annoyed by them. Experienced leaders welcome and manage the input of diverse personalities.

19.6.2 Have Clearly Articulated Goals

Leadership is all about knowing what needs to be done and how to get it done. Thus goal setting is a crucial element of great leadership.

We add that missing a milestone should not just be met with a shrug and a postponement of its due date to another day. Rather, missing a milestone should be a "call to action!" How can the overall targets of the project still be met? How can one ensure that there is not a flow-on effect that potentially endangers the overall project?

We also refer the reader back to the discussion on goal setting in Sects. 5.3 and 16.3. It might be helpful to reread those sections now. Ask yourself the relevance of the material as you approach a more senior phase of your career. Together with the experience you have now acquired, the previous discussion should yield further insights that will be evident from your new vantage point. In particular, we reiterate the following three points.

- First, the more complex, or uncertain, a goal is then, the more important it is to break it down into bits ("chunking") and scenarios. (Refer also to the discussion in Sect. 5.3.)
- Second, do not reward behaviour that has no benefit, i.e. do not add a bonus to early project completion if this is of no value; it is discouraging for people that have worked hard to speed things up. There is also the potential for quality or safety sacrifices being made; see Sect. 16.3.
- Third, make sure that you get an appropriate balance between competing goals, e.g. time- and cost-efficiency on the one hand versus the impact on quality, safety or legality; see Sect. 16.3.

19.6.3 Believe in What You Are Doing

An old adage is that if you do not believe in yourself then you can hardly expect others to believe in you. Of course, belief in one's self is not to be confused with arrogance. In other words, believing in one's self is manifested by quiet confidence rather than by trumpeting your importance from the highest hill.

> If you do not believe in what you are doing then contemplate finding something else to do.

19.6.4 Set Priorities

In all jobs, especially, at a senior level, there always seems to be too much to do. It is all too easy to "drown in" a sea of detail. Great leaders can see through the detail and set the appropriate priorities to get the job done.

19.6.5 Listen to Others

You will never be recognized as a great leader unless you are first a great listener.

19.6.6 Recognize Your Own Weakness

It is through recognizing our own strengths and weaknesses that we prosper. We have discussed this earlier and have encouraged you to recognizing your own strengths, see Sect. 19.6.3. Recognizing your limitations is equally important. Embracing differences and diversity as a source of creativity is yet another important leadership skill (see also Sect. 19.6.1).

19.6.7 Delegate

Closely related to the importance of recognizing one's weaknesses (see Sect. 19.6.6) is the skill to delegate.

If you cannot delegate then you certainly cannot run an entire university or industry. You will be limited by your personal capacity. No matter how large your capacity may be, it will always be less than if you leverage it with a team.

Stefan experienced a senior manager in a company that was an extreme example of an inability to delegate. He had an incredible work-ethic, regularly working from 7 am one day until 2 or 3 am the next day; home, rest, shower, back to work. Until this work-ethic got him promoted to an extremely responsible position where the tasks were such that no single person, no matter how brilliant, could handle them without delegation—which he refused.

So, inevitably, he started slipping behind, missing deadlines. All of this he would try to counter with, yes, you guessed it, yet more working hours. Now, of course, he was also beginning to pay a price with worsening health, upset family and general burnout. The saddest thing, perhaps, was that he became very bitter, feeling unappreciated for all his work. "Nobody", he said, "nobody could work harder than I do".

True, maybe, nobody could have worked harder—but definitely more effectively, by delegating better, by working smarter (see also Sect. 23.4, 23.5). Make no mistake, this is not only a junior manager issue, this is a problem that senior managers struggle with as well!

19.6.8 Understand Teamwork

As a follow-up to delegation, it is important to understand the value of teamwork. It is as true in industry and universities as it is on the playing field.

> Remember there is no "i" in the word "team".

19.6.9 Inspire

A great leader has the capacity to inspire. Great work always follows when people are inspired.

19.7 Summary of Leadership Qualities

Great leaders
1. Celebrate successes (of others not themselves!)
2. Are open and communicative.
3. Inspire risk taking by setting an example.
4. Are candid, open and trustworthy.
5. Make unpopular decisions when necessary.
6. Build confidence in others and avoid destructive criticism especially when a calculated risk has been taken.
7. Know how to delegate.
8. Know how to build a team.
9. Take every opportunity to praise the good outcomes achieved by others either by a quick email or, even better, by visiting the person's work space.
10. Never claim other people's successes as their own.
11. Bounce back with enthusiasm following a temporary setback.
12. Make all subordinates feel special and appreciated, be it the CTO or the cleaner.
13. Tell people forthrightly when both the organization and their own interests are best served by them changing their approach.
14. Never "kiss-up" and "kick-down". That is the realm of psychopaths!
15. Never promote endless "new visions" especially if either (a) the current one is already working well or (b) the current one has yet to take hold.
16. Always acknowledge problems quickly and move to remedy them rather than to obfuscate them.

17. Learn from mistakes (otherwise why on earth would you make them in the first place).
18. Match resources to expectations.
19. Say as little as necessary to communicate the core idea or vision.
20. Become the key protagonist for your organization.
21. Operate an open door policy (within reasonable constraints on time).
22. Do not blame failures on the legacy left by earlier holders of their position.

19.8 Learning from Other Leaders

We strongly urge our readers to learn as much as they can by listening to others. Many books are available on the lives of great leaders. We suggest that it is a good idea to read some of these.

19.9 An Example of Great Leadership

Graham recently asked a very senior engineer from a large company if he could define what it took to be a great leader. He replied "It's straightforward".

> "A great leader is simply a person that other people want to follow!"

He said that he had dealt with thousands of engineers during his career but singled out one to exemplify his claim. He said that the person was technically extremely competent but added that hundreds of others were similarly qualified. What distinguished the person from all of the others was the fact that he understood, nurtured, developed and mentored those people who reported to him. In turn, his team was prepared to follow him "no matter how arduous the journey". In other words, this person was the quintessential example of a great leader!

19.10 Judging Others

An inevitable consequence of your transition to more senior roles is that you will need to pass judgement on others. It is in this role that you can exercise true leadership. Clearly "quality" should trump "quantity". If you are in a senior academic position then it may be informative to revisit the comments we made regarding quantity versus quality in Sect. 15.4.

19.11 Summary

Part IV of the book moves on to times when you approach a more senior appointment, be it in the academic, private, government or consulting sector.

This current chapter picks up where Part III ended: the cycle of success. In a senior position the cycle consists of the same basic elements already described: position plus reputation, resources plus skills and leadership. However, the content, weight, leverage and intensity of the elements increase as your career matures.

In the academic sector you might find your position advancing from being head of a research group, to becoming head of department and possibly dean or president. Equivalently, a career in the private sector might grow along the lines of head of department, head of a division and ultimately to CEO (Sect. 19.2).

In these positions your reputation is even more important as you are more exposed. However, now you will also be held accountable for the reputation of other members in your team. A breach of ethics does not only impact on the violator, but also on you as their leader. It is therefore important that you share your values with your team at an early stage. If an incident does occur, think carefully about how you manage it. People incurring reputation damage often do not only do so from the original incident, but from their handling of it.

Parallel to the growth of your position, you are likely to gain access to larger resources and skills: you will probably attract larger budgets and be able to draw on the skills of a large team (Sect. 19.3). So the humble beginnings of growing your cycle of success in your junior years (Chap. 18) can grow to substantial proportions in your more senior years.

We now clearly see, that the cycle of success can go in two directions. In the positive direction, your growing access to resources allows you to deliver ever-greater results, a fact that in turn grows your reputation and continues the cycle. In the negative direction, your results may fall short of the resources you have been given and promises you have made, damaging your reputation even more catastrophically than would have occurred had you occupied a more junior position.

We therefore draw attention to the importance of delivering on promises (Sect. 19.4) and conducting efficient meetings (Sect. 19.5) as you lead larger teams and command extra resources. For example, if you waste an hour as a student you have done just that: you have wasted one hour. If you gather 8 people in a meeting as a senior, then every hour translates into the equivalent of a whole day—precious time not to be wasted. If you have 40 people in a meeting, every hour translates into the equivalent of a whole week!

Sections 19.6–19.10 are devoted to casting the third element of the cycle, leadership, into the context of a more senior position. Many of the topics will ring familiar to you from your actions in more junior roles: appointing the best people, articulating clear goals, setting priorities, listening, recognizing your weaknesses, delegating and inspiring. You will find a summary of desirable leadership qualities in Sect. 19.7. However, as with the other elements of the cycle, these aspects carry more weight in more senior positions.

19.12 Further Reading

[1] EA-Engineers Australia "Top 100, Australia's most influential engineers" Vol. 86, No. 6, June 2014, pp. 45–75.

Chapter 20
Public Speaking and Dealing with the Media

20.1 Overview

This chapter discusses communication skills with particular emphasis on the role of senior people.

20.2 Public Speaking

As your career advances, you will increasingly be offered the chance to address ever-larger audiences. This will often be where the "public image" of your organization is placed on display for all to judge. These are great opportunities. Therefore, it is well worth training yourself to be comfortable with public speaking: You want to neither shy away from the offer (missed opportunity) nor deliver a poor presentation (botched opportunity).

As a specific example, Graham began his academic career addressing 20–30 people in the form of standard conference presentations. As his academic career advanced, he moved to larger audiences. On several occasions, he addressed several thousand listeners when he was asked to be a keynote speaker at major international conferences.

Irrespective of the size of the audience, there are some simple rules that will help you give an effective and positive speech. Firstly, we remind you of the recommendations of Toastmasters:

1. Tell them what you plan to say.
2. Say it.
3. Tell them what you have said.

These are basically three different perspectives on your subject. The first perspective should capture the audience's attention and inspire their interest, put them all on the "same page" and provide an overview and road map of what has to come.

The second perspective presents the main body, namely the substance, together with supporting evidence and arguments of your talk. The final perspective distills the talk into the take-away message(s) and puts it into some larger context, such as the value of the presented topic for the future or open questions needing attention.

You should be aware that you have only 30 s to capture the attention of your audience. Thus, get into your theme quickly.

Things you should not include:

- Never start with an insecure apology ("I am actually a poor speaker..." - and then go on to prove it!.
- Never spend a long time thanking others for their "kind invitation".
- Be cautious about telling a joke that you feel is funny. It may waste time and, worse, may offend some in your audience.
- Never cover a "smorgasbord", of ideas, i.e., be careful not to have a set of ideas that is too wide and unrelated.
- Do not use visual aids as a "crutch".
- Do not build expectations in the audience without delivering on them.
- Do not have a long-winded introduction - get into your theme!

Things you should include:

- Plan the length of your talk - should it be three minutes, five minutes or 50 min.
- Have a clear message in your talk.
- Match the talk to the expertise and background of the audience. Should your talk be narrow or broad? Does your audience include non-professionals?
- Think about what you hope to achieve with your presentation.
- Ask yourself, what is the "take-home message".
- Capture your audience's attention or "go home".
- Think about what might excite your audience rather than what excites you.
- Judiciously blend "analytical presentation" and "story-telling" styles - they appeal differently. The former spells things out; the latter evokes an experience in the listener. Craftful blending can be very powerful.
- Do not forget the effectiveness of a pause.
- Give your audience a good reason to listen to you.
- Remember to be clear and articulate. "Speak up" or "Shut up".
- Talk "to" your audience, not "at" your audience.
- Remember you want your audience (be it 10 people or 3,000 people) to feel you are talking directly to them as individuals.

We have repeatedly mentioned the importance of being familiar with your audience and their backgrounds (lay people, students, experts etc.).

Ideally, you should learn about the background of your audience in advance so you can tailor your speech accordingly. However, with a big audience, you sometimes need to adapt. As an example, Stefan once witnessed a well-known theoretical physicist and impressively demonstrated the skill of adaptation. The event was a podium-discussion on Quantum Mechanics, Relativity and String Theory.

The first two theoreticians opened with their views - delivered in style, language and terminology that physicists would use daily and would be very comfortable with.

The third speaker opened his contribution with: "Let me ask for a quick show of hands, please: how many of you are very familiar with Quantum Mechanics and Relativity?" And then: "And how many of you do not deal with them on a daily basis"?

It turned out that the vast majority of the audience were in the latter category. Indeed, many were lay people.

So, the speaker then tailored his address to his audience and said something like: "Well basically Quantum Physics has been tested many times on really small stuff and it works extremely well at this scale. Also, Relativity has been tested many times on really large stuff and it works extremely well for these problems including planets and galaxies. But, here is the problem: When you try to merge the theories to get one paradigm that works for both small and large stuff, -then it turns out that they are mathematically incompatible. But how can this be? Two theories, each well proven in practice, but mutually incompatible? Trying to resolve this incompatibility is what String Theory attempts to do".

We can assume, of course, that the speaker had an alternative, more technical, opening statement at hand to use if the initial show of hands had revealed a more expert audience.

A beautiful display of command of subject as well as public speaking skills.

20.3   Dealing with the Media

Senior people in industry and academia need to be prepared to deal with the media. Here your audience will be "indirect" rather that "direct". In other words, you may only be talking to a single interviewer but when your discussion "goes to air" it will be heard by thousands of people. Making such a presentation takes special skills. If you need to do it frequently or if you might be called upon to comment on current news such as breakthroughs or crises, then we advise that you consider taking a specific media - training course.

> Key issues when being interviewed are:
> - Think carefully about the message that you want to deliver.
> - Package your ideas in "media friendly" grabs.
> - Do not answer questions with a simple "yes" or "no", better to say, "What I can say is that..."

In the author's experience, dealing with the media can be one of the hardest things that you do. The media always want catchy phrases that "sell" or which create controversy. However, you should be careful not to be trapped into making exaggerated claims or allowing the media to quote you in such a fashion as to create a controversy or stir up public opinion against you or your organization.

20.4 Summary

This chapter looks at two aspects of communication you might come across in a senior position: public speaking and addressing the media.

The communication and speaking skills you have developed in your early professional years are a solid investment to build on in senior positions. Many of the principles remain the same (Sect. 20.2). Having internalized the fundamentals as second nature allows you to concentrate on two new issues that arise specifically in senior positions.

Firstly, more is at stake. This is equally true for the opportunity and the risk. Addressing larger audiences on more public platforms is a great opportunity. On the other hand, standing on the same platform with a poorly prepared speech or an ill-advised statement will be more visible and you will be less forgiven. So, if you prepared well in your junior positions, prepare extra well in your senior positions!

Secondly, the background of your audience might be more diverse. It is one learning curve to address ever-larger audience of similar background, say, a group of a dozen students in a tutorial versus a lecture hall with several hundred students. Similarly addressing a small department in your company is quite different to addressing the entire company.

It is an additional learning curve to address a public audience of vastly different backgrounds, covering a spectrum of experts and lay people. The more you know about your audience, the better you can address the opportunity. The better you master your subject, the faster you can adapt as you get to know your audience (see Sect. 20.2).

When it comes to addressing the media, nothing beats formal media training. If it is likely that your position will require media interaction, we strongly recommend you invest in such training. The more difficult situation is if you are suddenly cast into the media spotlight unprepared. This can be for either positive or negative reasons, - both can be quite intricate to manage. Section 20.3 draws your attention to this matter so you can decide whether, and when, media training is appropriate for you.

Chapter 21
Job and Career Changes

21.1 Overview

One of the exciting things about life is that it is forever changing with new adventures and new challenges.

Indeed, achieving success in any endeavour often comes down to our capacity to adapt (see Ref. [1], Further Reading).

In this chapter, we address issues that arise when you contemplate a job or career change in more senior positions.

21.2 Embracing Unsolicited Change

Unsolicited change is something that everybody will need to deal with at sometime or another. Unsolicited change can arise due to evolutions in technology, downsizing, health issues, changing markets, etc. These changes are equally relevant in the university and private sector. A related issue is changing attitudes to the way that excellence is measured which can "wrong foot" an individual even though they were previously judged to be performing at the top of their field.

Part of the art of dealing with unsolicited change is to know when to resist it and when to embrace it. An example of resisting change is when your group is under serious external attack but you decide to fight on and win! Alternatively, this could be the very signal to do something entirely different. This is nicely captured in the well-known Serenity Prayer (see Refs. [2, 3], Further Reading).

> "God, grant me the serenity to accept the things I cannot change, courage to change the things I can, and wisdom to know the difference".

It is difficult to give generic advice but we would like to point to some important factors in dealing with forced change. Some helpful suggestions are:

> - Watch out that fear of the unknown does not force you to become immobile.
> - Be mindful that fear shuts down your creativity so you may loose the capacity to recognize the need for change.
> - The human body's stress system does not distinguish between imagined crisis and a real one. Focusing on negatives has one certain outcome - you will experience stress!

Conventional wisdom is that it is "natural" to be fearful and resist change. However, this is not true. It is a good exercise to think of all those times in your life when you longed for change. What is true is that exploring change is best done from an environment that feels nurturing and safe.

> If you have to confront others with change, then it is important to create an environment in which change is viewed as a positive rather than a negative step.

We all know that we should not give up after our first failure. The extent to which this is true is easily forgotten. So, after several attempts you may feel it is time to move on. However, often multiple failures are natural. If you think about how many times a toddler stumbles before walking then you will know that eventual success often follows many failed attempts.

Change may contain great opportunities to move into new areas or to improve your work environment. We advise that you should seize the moment and be courageous.

> A provocative, but useful question to ask yourself in the face of a crisis is "why is this crisis the perfect opportunity for me to improve my situation?"

21.3 Should You Remain Technically Active or Move into Management?

The question set in the title of this section is by no means trivial. The answer depends a lot on your own skill and ambition. It is fair to say that many people gravitate to administrative positions as their careers develop. However, this need not apply to you.

21.3 Should You Remain Technically Active or Move into Management?

Moving from a technical to management stream is often a clear career choice with "no way back". However, it is possible to "keep a foot" in both camps. For example, Graham was, at one time, both Dean of a Faculty of Engineering as well as Director of a Special Research Centre. Indeed, we have noticed that many Deans of Engineering throughout the world are able to maintain healthy research careers.

However, you should keep in mind that senior management as well as senior research positions are becoming increasingly demanding and thus it may be necessary to make a clear choice.

Professor Arie Feuer had the following comments about change:

"A common experience in academia is that at some point you feel that you have exhausted the particular area you have been researching for many years. At this point some choose to turn to university managerial positions - Dean, Vice President of the University or other such positions. Others choose to change their area of research. Some decide to leave academia and move to industry, possibly tempted by financial rewards. Finally, there are a small number who, after retiring from academia, choose to get involved in industry typically in a small start-up company. This can be particularly rewarding as for many engineers the actual 'proof of a pudding lies in the eating'."

21.4 Continuity and Leaps into the Unknown

Changes can either be continuous or, at other times, more of a leap into unknown territory. Such leaps can lead you into related lines of activity such as creating a spin-off company or they can be a leap into a totally different area of work.

> The main thing one learns from doing a doctorate is that all problems, no matter how difficult, can be solved by careful and meticulous attention to detail.

In this context, we can relate the story of Adrian Medioli, one of Graham's doctorate students. He began his career in industry designing and commissioning control systems. However, after 7 years, he was looking for greater personal fulfilment. So, he left to become a doctorate student. On completing his doctorate, he became a full-time researcher. He worked on three principle projects:

- Alarm root cause analysis in industry
- Ambulance scheduling
- Developing an artificial pancreas for Type 1 diabetes patients

He holds a senior leadership role in the latter position.
Our final recommendations are as follows:

> Follow your passion and natural inclination. If this happens to lead you to make large changes, then you should embrace the opportunity. On the other hand, if you are naturally more conservative, and the changing world forces change upon you, then take heart as some change is always inevitable and can be rewarding and beautiful.

21.5 Summary

This chapter addresses career changes that fall into two broad categories: unsolicited change, such as being laid-off, retrenched or fired (Sect. 21.2); and voluntary change such as changing your role (e.g., moving from a technical or operations position to a management position (Sect. 21.3)).

The more you embrace the change, the better you will transform.

This raises two questions: first, how should you embrace change that you do not want (loosing a job); and second, how do you pursue, or deal with, opportunities of change if you feel undecided (moving from a technical area to manager or changing employment sector).

When it comes to forced change (Sect. 21.2), we draw your attention to challenge conventional wisdom that claims: "It is human nature to be fearful of change". We advise that you think positively and recall all of the times that you have successfully navigated change - we all have.

The question of voluntary change, such as from technical expert to manager or from academic to private sector, is riddled by possible difficulties and self-doubt: how will you decide, and will you regret? Section 21.3 offers some pointers.

21.6 Further Reading

[1] Max McKeown, "Adaptability: The Art of Winning in an Age of Uncertainty" Kogan Page Limited 2012, Reprinted 2013, United Kingdom and USA.
[2] F.R. Shapiro, "Who wrote the Serenity Prayer?" Yale Alumni Magazine, July/August, 2008.
[3] R. Niebuhr, "The essential Reinhold Niebuhr: Selected essays and addresses, ed: R. Brown" Yale University Press, September, 1987.

Chapter 22
Mentoring and Succession Planning

22.1 Overview

In this chapter, we discuss the role of mentoring in a senior position. We also briefly address the topic of succession planning.

22.2 Mentoring

Previously in the book we have addressed the importance and rewards of mentoring when you hold a relatively junior position. Here, we address mentoring at the senior level. Amongst other things, your senior position may mean that you will now be mentoring other mentors. This kind of mentoring brings the same benefits as mentoring at a more junior level.

> Mentoring:
> - Is a fulfilling experience.
> - Builds your reputation and breadth.
> - Grows networks.
> - Helps you keep you up-to-date.
> - Helps grow a team of future talent.
> - Allows you to inject your experience at lower levels in the organization.
> - Is a wonderful mechanism for building lasting relationships.

To benefit from all of this you will need to augment your junior mentoring skills by senior mentoring skills. The extra dimensions needed at the senior level include mentoring in nontechnical areas including life skills and offering leadership qualities.

Also note that, at a senior level, you may well be mentoring people who work in technical areas that lie outside your own specific technical expertise.

> Your key task is to provide an environment in which the person being mentored can grow. This includes giving them room to develop their own personal style and strengths.

> As a mentor, you should encourage growth and diversity.

We specifically mention this to avoid the inclination of some managers to evangelically preach what worked for them. Remember that these are rapidly changing times so the details of what worked for you might not work for the next generation. So, we suggest that you focus on time-less principles, whilst still providing appropriate illustrations from your personal life.

22.3 Succession Planning

An issue tightly linked to that of mentoring is succession planning. Many people associate succession planning only with their own retirement. However, there are many reasons why you may need a successor at other stages in your career, such as

- You receive an upwards promotion.
- You leave the organization or country.
- You disengage from a particular activity, e.g. you move away from a project.
- There is always the possibility of an illness or other unexpected event. (The "runaway bus" scenario.)

> The advice that we offer is
>
> (a) Never make your potential choice of successor known too early.
> (b) Make sure you evaluate your successors potential according to the *future* job requirements rather than the person's performance in their current role.

The reasons for not making the potential successor known too early include:

- It can be a terrible blow to the person if it does not work out - especially if you are the person who has both told the person and subsequently decided on another choice. Alternatively it may turn out to be "not your call".
- There is a risk that other potential candidates are disenfranchised.

- There is a risk that others will try to ingratiate themselves with the chosen successor.
- You run the risk of becoming the "lame duck" and hence losing your own authority.

On the point concerning evaluating a person based on the future job requirements rather then the current, this can help avoid the well-known Peter's Principle, i.e. "Everybody rises to their own level of incompetence" (see Ref. [1], Further Reading). Alas, it is all too common that people who perform extremely well at a particular level are promoted to positions where they are "out of their depth". This does nobody a service, especially not the person in question.

Always give your successor space to put their own mark on your previous position. Do not try to upstage your successor unless you really do want to create a lifelong enemy!

22.4 Retirement

> If you are retiring then there is the issue of how to fill a sudden void in your own life. We believe it is very worthwhile to do so conscientiously and proactively.

Actually, Graham is, at the time of writing this book, "semi-retired". Nonetheless, he continues to work extremely hard (more than 50 h a week) on his research. Friends form outside the academic circle question why he would do this.

Graham's reply is simply:

> "Retirement is about choice and I choose to continue to engage in research".

Indeed it is interesting to note that many academics continue to do research in their post-retirement years. Actually this is a very positive acclamation of the incredible "pull of research". It gives the most powerful argument we can think of for undertaking a doctorate in the first place!

22.5 Summary

Important aspects of the job of senior people are mentoring and succession planning.

Duty, threat or joy?

Section 22.2 highlights the joy of mentoring. You may well have mentored many peers in your junior years. Equally, you are very likely to have been mentored by others. As you move into more senior positions the importance of mentoring remains but it may change in detail. For example, you may now be mentoring other mentors.

A topic related to mentoring is succession planning (Sect. 22.3). Succession planning is often emotionally loaded, as it can trigger fears of being replaced or being out-classed. Whilst these are powerful concerns, successful leaders always continuously grow their skills to manage succession planning as an environment of growth instead of a Petri dish of toxicity (Sect. 22.3).

22.6 Further Reading

[1] L.J. Peter and R. Hull, "The Peter principle: Why things always go wrong" William Morrow and Company, 1969.

Chapter 23
Work-Life Balance

23.1 Overview

We first covered the topic of creating an appropriate work-life balance that works for you in Part I of this book. At that point we said that this was a life-long issue. It is this life-long relevance that motivates us to revisit the topic in this, the final, part of the book.

23.2 Reflecting on Work-Life Balance

If you fail to get your work-life balance in harmony, then sooner or later your body is likely to "stop you in your tracks". There is an abundance of people, who have overworked themselves and ended up with a break from work imposed on them through burn-out, accident or disease. Later, we will say something about how to use such time proactively, if it happens to occur. However, getting your work-life balance right in the first place can significantly reduce the probability of this body-driven intervention being necessary.

In his coaching business, Stefan sees many clients who come for specific advice on work-life balance. He always starts by exploring two questions, which are often overlooked and which can seem somewhat threatening when they are first asked:

> (a) Do you truly want a different work-life balance?
> (b) Is your current personal life such that you feel positive about giving it greater emphasis?

Regarding point (a), there are people who gain such high satisfaction from their work that they are completely fulfilled. In this case, we advise:

Fig. 23.1 "Think about how you will feel on your death bed. Will your present priorities and choices regarding your work-life balance remain valid, without regrets, when all else is gone?" - see Sect. 23.2

- Manage your self as best you can so as to avoid "burn-out".
- Do not stress about attaining a different work-life balance - maybe it isn't for you!
- Watch out for "collateral damage", i.e. make sure your fixation on work is not damaging others such as your partner, your children or even your work colleagues.
- Every so often, say every five years, reconsider your work-life balance.

Think about how you will feel on your death bed. Will your present priorities and choices regarding your work-life balance remain valid, without regrets, when everything else is gone (Fig. 23.1)?

Part of the motivation for this book arose from Graham's interactions with John Edwards, a colleague from industry. John became very ill in his later life and one of his key messages to Graham during their Friday meetings was

> Do not put off until tomorrow what you dreamed about yesterday.

John advised Graham to go to Antarctica (somewhere John had gone before his illness took complete control). Graham acknowledges that this was amongst the best advice he ever received.

> Message to reader: Plan your version of a "trip to Antarctica" now!

It is surprising how many people fail to get their work-life balance right because they don't like what the second half of the balance has to offer them. If this applies to you, then you should think about your relationships, your living situation or whatever awaits you outside of work. In this case, it is futile to work on time management when the real problem lies elsewhere.

Whilst it can be quite threatening to address the real issue, it is helpful to use one of our previous strategies and explore what you are saying "no" to if you give into not changing your situation. You could be saying no to a satisfactory personal life and risk collateral damage and regrets.

23.3 Reacting Positively to a Forced Time Out

If your work-life balance tilts against your health, you may well suffer burn-out. Indeed, this happened at one point in time to both authors. Under these conditions, we suggest that you use the forced "time-out" wisely. For example, Stefan, during a 2 month period of recovery, found it helpful to reflect on three questions, courtesy his colleague Tamas Mayer

- What do I miss about work?
- What do I not miss about work?
- What do I learn from this?

His answers to these questions led him to make significant changes to his life.

23.4 Working Smarter Not Harder

If you have read the previous sections and you have convinced yourself that you really want to change your work-life balance, then this section will provide guidance as to how you can achieve this.

As your cycle of success and your work starts impinging on your work-life balance, then a natural temptation might be to work longer and harder. If this is a virtue and if you master the skills, it will still only go so far. So, by all means, work harder and longer but note that working smarter will ultimately be the best way forward.

As with previous topics, there have been many books written on this topic, so we will simply highlight a few issues

- Prioritize your work in advance. Recall that when you get stressed then the last thing you are likely to do is step back and take stock of where the pressure points lie and how they can be addressed.
- You should set down in writing the pressure points on a white board or flip chart, so you *see* the "big picture". The reason this works is that doing this frees up your brain capacity and enables the visual brain to take over and allows you to prioritize.

23.4.1 Reduce Task Swapping

Parallel tasking and task swapping may seem like a good idea but it actually consumes energy and time so it is often more efficient to group tasks than to swap too rapidly between tasks. Part of this is to avoid being electronically connected at all times. Instead, set aside a specific time in which you deal with electronic communications. Be mindful of your own capacity to effectively swap between tasks and remember that irrespective of your skill level in this sphere, rapid task swapping is generally less efficient then focusing on one task at a time.

The trick is to find the right "dwell time" for tasks. Then use the time efficiently, moving on only when a fresh task is necessary to reinvigorate you.

23.4.2 Task Completion

Avoid the habit of having a long task list of jobs that are 90% completed. All too often, the joy of having one job finished gives into the pressure of starting the next job.

> It is extremely empowering to finish a task and hence have it irrevocably removed from the "to do" list.

23.5 Energy Management

We also want to point out the advantages of shifting your perspective from "time management" to "energy management". Obviously careful management of time is very important. However,

> An over emphasis on "time" has the following two limitations:
> - It gives excessive emphasis to the 24 h day whereas energy has no associated time-scale.
> - It drives you to work ever more and opens the pitfall of overlooking efficiency.

Both these difficulties are addressed if you extend your skill of time management to encompass energy management. As a simple example, energy management would encourage you to take a break, rejuvenate, or eat a piece of fruit whereas an overemphasis on time might lead you to believe that such recharging of energy stores are nothing more than a waste of time. It is exactly the opposite when viewed from the perspective of energy management.

> A shift to energy management will lead you to be more subtle about how you utilize your time.

Finally, it is amazing how often that an answer to a difficult problem or clarity on our position occurs exactly when we take a break from worrying about it. This could be during a period at the gym, relaxing on the beach or your own version of "Antarctica".

As a specific example, Graham recalls being stuck on a particular research problem for almost a year. He decided to take a few days off and relax by the sea. Bingo! The problem was solved. You, the reader, should not be surprised. This is a very common occurrence.

23.6 Delegating

An important aspect of achieving an appropriate work-life balance is your ability to delegate.

> You need to ask, "does a job really need doing, and if so, do I really need to do it"?

If you are unable to delegate, then your capacity will always be limited by what you can achieve alone. Incidentally, the same is true if you delegate but then micro-manage the delegated task. On the other hand, if you learn to delegate properly, then your capacity for ever larger projects is virtually unlimited. Of course, this presupposes that you continuously hone your delegation skills.

We have warned about the dangers of micro-managing delegations. Of course, this does not mean that one should be "laissez-faire". The latter neither provides leadership nor appropriate accountability.

23.7 Summary

The more your cycle of success grows, the more "weight" will pile up on your shoulders. At times this is rewarding as it creates stimulation, challenge and inspiration. At some point, however, it might confront you with adverse effects, compromising your health, social or family life.

Section 23.2 reflects on work-life balance issues. One of them is: "Why am I struggling to balance my professional and private life? In particular, the time and energy I have for either"?

Beyond the obvious answer of: "I just have too much to do in the available 24 h day, too much responsibility". This section encourages some further reflection.

For example, maybe 70 h per week at work "is simply you"? Maybe you are so driven by your work that this is the "balance" that suits you? It can be quite daunting to face some of the questions posed in Sect. 23.2. We suggest, however, that giving yourself permission to honestly ask them of yourself is well worth the effort before you attempt to change your work-life balance just "because one is supposed to".

Sometimes, a radical change in work-life habits is forced upon us by illness, accident or lay-off. Gentler occasions for time-out are vacations. Section 23.3 provides three helpful questions to ask yourself, during recreational or forced time-out: What am I missing, what am I not missing, what do I learn from this?

Sections 23.2 and 23.3 are intended to alert you to the fact that it is often difficult to change your work-life habits. Maybe you feel that your present habits are working well for you or, conversely, that changing them would not work well either!

23.7 Summary

The remaining sections of the chapter address managing your work efficiency. This is about *how* you do, not just *what* you do.

In this vein, Sect. 23.4 is about working smarter not harder. Section 23.5 recasts thinking of time management into energy management. The creative potential for this is that, time is unmercifully rigid whereas the energy and efficiency you fill it with is up to you.

23.8 Further Reading

[1] D. Rock, "Your Brain at Work: Strategies for Overcoming Distractions, Regaining Focus, and Working Smarter All Day Long", Harper Business, 2009.

Chapter 24
Keeping the Bigger Picture in Focus

24.1 Overview

In this the final chapter we examine issues regarding keeping the bigger picture in focus.

In earlier chapters, we have discussed extensively about getting your cycle of success spinning. Now, we arrive at the final question, namely, where do you want this wheel of success to finally lead you.

There may be digressions along the way but a long-term vision sets the path.

Making a success of difficult tasks is often fed by dedication and long-term planning.

To quote the Physicist, Paul Dirac

> "You cannot realize your dreams unless you have one to begin with".

24.2 Keeping the Bigger Picture in Focus

> In contemplating your goals we would advise to set a long-run vision that is bold. If you set a mediocre goal, then chances are that you will achieve it!

We therefore believe that you need to be bold, provided of course that you balance this with tangible, reachable steps in the short term. As your career develops and you accomplish your short-term goals, then you will need to raise your long-term goals commensurately. Since you own your goals, then mid-term corrections are perfectly valid since you will know that those goals have been inspired by boldness.

It is helpful when setting goals to be very specific. This often includes tangibles such as achieving a certain position, a personal achievement, a particular invention, a certain income, a lifestyle, some specific recognition or an award. However,

> It is important to keep in mind that satisfaction is a "feeling" as is happiness and contentment. You may therefore need to augment your long-term vision with an overarching umbrella of feelings.

The core intent of tangible goals is to deliver on your feelings. Being explicit about this link will help you to adapt your goals if and when they do not deliver the feelings that you aspire to achieve.

You definitely do not want to spend your life doing something that you ultimately realize was futile. (This is beautifully articulated in Ref. [2], Further Reading.)

We also recommend that you extend your bold goals by the values that are important to you and by which you want to live. These values could include being ethical, compassionate, generous and kind. Oddly, it may even help if you think what others might write as your epitaph. Would you be pleased to hear their summary of your life and its value?

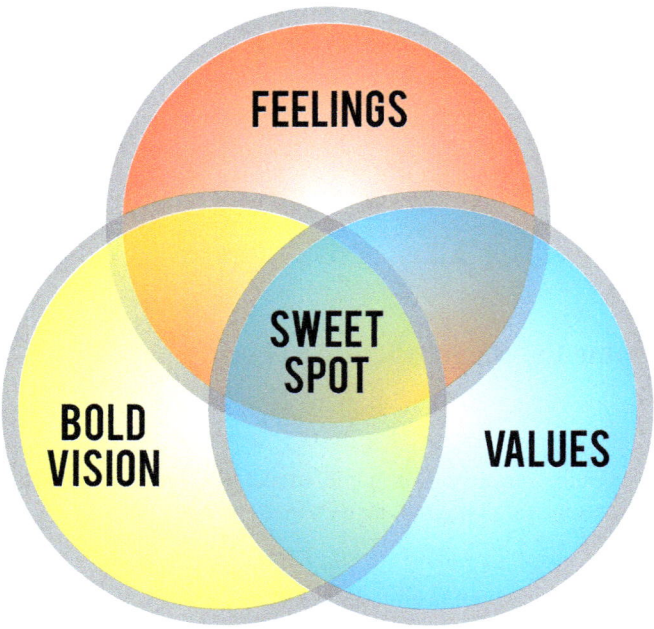

Fig. 24.1 The broader picture: Overlapping vision, values and feelings

24.3 Regaining Balance When Injustice Strikes

You should strive to achieve your short-term goals, at least, or learn from not achieving them. You should occasionally revise the bold vision and the feelings and values that are important to you. As you approach goals make sure you turn them into milestones so that you open the door to the journey beyond the goal (Fig. 24.1).

The "sweet spot" is where goals are aligned with your values and feelings such as a sense of fulfillment, satisfaction and passion.

24.3 Regaining Balance When Injustice Strikes

For every honour ever bestowed, for every achievement ever celebrated, for every award ever won, there exists an alternative group of people who believe they were the rightful recipients.

In addition to famous examples, the authors personally know of tens of people who have lived decades consumed by bitterness because they missed out on recognition or an honour that was bestowed on somebody else. Understandably, having expended tremendous energy on a goal that is attributed to somebody else, it is easy to flaunt the injustice and try to correct it.

We certainly advocate that you calmly argue your case. However, there comes a time when it is wise to accept that (perceived) injustice has occurred and move on! Otherwise you run the risk of turning an injustice into a major catastrophe *for you*.

In reality, an injustice may not be as clear cut as one may believe, it is in the midst of an evolving situation.

For the sake of argument let us assume that you strongly believe that an event is grossly unjust towards you. Let us also assume that you have exhausted whatever appeal mechanisms you may have but, the decision stands.

In this case you have to live with it. So now the question becomes, how *best* do you live with it? The general answer is that when you can neither prevent nor overturn an injustice against you, you can still craft your attitude towards living with it.

Obviously, one possible attitude is to spend the next decades consumed by bitterness and anger as the victim of injustice. Particularly when it comes to the most prestigious and hard-earned prizes this definitely happens. As mentioned, both authors personally know such cases.

If you choose this attitude, however, it is unfortunately *you* who will pay the greatest price: the injustice of the lost prize is still felt but now you add to it decades of bitterness. It might be tempting to blame the decision makers for your bitterness - after all it is they who meted out the injustice in the first place! Alas, however true that may or may not be, it is still you paying the price by living in bitterness, caught up in the past.

It is therefore our advice to embrace a damage containing attitude for dealing with the situation. You can contain the impact of the injustice by not compounding it. Depending on the severity of your perceived injustice, this can be easier said than done.

Stefan and his colleague, Chomba Hermanus, have found the following thought helpful: "Grieve how bad the injustice was, but do not let it fester or allow it to become worse by contaminating your future."

We believe that it is well worth trying the above advice! It is natural, even important, that a grieving period might have to pass before you can tackle what amounts to a forgiving process.

Many people confuse forgiving with condoning or down-playing an injustice. They are not the same at all: you can forgive and still *not* condone, just as, you can forgive and still seek legal action or other compensation.

Forgiving is much less for the one being forgiven, it is rather more for the one doing the forgiving: they come to a closure, let the negativity be as large as it was, but contain it in the past so it does not contaminate the rest of their life. In lieu of having won the initial prize to begin with, this is certainly a terrific second prize for which to strive!

If you are in the fortunate, and rare, position that this has never happened to you, then it is helpful to think through the issues in advance so that you have a clear vision of how to cope should it eventually occur.

24.4 Broader Life Issues

The advice given in Sect. 24.3 applies more broadly in life. We recommend the book written by Rosie Batty (see Ref. [1], Further Reading). Rosie is a truly inspirational person. She is a domestic violence campaigner who was recognized as the 2015 Australian of the year.

Her commitment to the domestic violence issue began in 2014 after her 11 year old son, Luke, was tragically murdered by his father. There are many similar stories of people who were able to rise above an overwhelming tragedy and turn it into a positive goal of benefit to others.

24.5 Embracing Serendipity

Whilst having goals, vision and planning are crucial to a successful career, there is always an additional unplanned force, namely serendipity. This includes coincidences, accidental discoveries and good fortune.

> Every great career and every great achievement depends to some degree on serendipity.

You may wish to think about your own life and see how pursuing goals and serendipity have each played a role.

> To embrace serendipity it is necessary to be prepared for the unexpected and then to capitalize on the opportunity when it arises. Here the crucial skill is to balance long-term vision with the capacity to capture the benefits of unplanned luck when it inevitably comes your way.

24.6 Our Final Wish for You

We wish you the skills to lead a career and life that brings you fulfillment, satisfaction and happiness. As the famous lyrics of the band Metallica go: "Nothing else matters!"

24.7 Summary

We have reached the final chapter of the book. It revisits a number of fundamental issues that were addressed in earlier parts of the book: goals, dealing and growing with rejection, as well as embracing serendipity.

The title of Sect. 24.2 combines the words goals, values and feelings. The importance of tangible goals has been a recurring theme through-out the book. However, here is an important extra point. Imagine you achieved a goal and it left you *feeling* void and cold. Now compare that experience to achieving a goal that left you *feeling* satisfied, happy and content. Which goal would you rather achieve?

We conclude that it is not about achieving a particular goal that is the most important thing but about achieving goals that are meaningful to you.

The importance of values is also easily overlooked, in particular for your bold goals. "Feelings" and "values" probably seem like a luxury in our goal-driven world. However, we should turn the perspective around. How sad would it be if you looked at an achievement only to admit: "I achieved the goal but it left me feeling void. Moreover, to achieve the goal, I compromised my values"?

That is why the title of Sect. 24.2 includes the words goals, values and feelings. We hope you may indeed be fortunate and achieve your goals. But if not, being able to say: "I was fascinated by the quest (whatever it might be), I devoted myself to it, and I did my very best whilst living by my values" is a very satisfying position to achieve.

A different challenge arises when you feel that you are not awarded the recognition you deserve: Sect. 24.3 looks at regaining balance after injustice strikes.

This is not a theoretical issue. For every award or honour that was ever won by somebody, there was always a group of people that felt they were the rightful recipients. If you should find yourself in that position, by all means, argue your case and use all available avenues for appeal. But if that fails, if you are in a situation where you feel injustice has struck and you have exhausted your appeal mechanisms, then your focus should shift towards your attitude of living with it. The more you can manage to contain the disappointment to an event in your past, the less you are likely to magnify its contaminating effect on your future.

The theme is broadened to dealing with overwhelming life tragedies. Throughout the book we have talked much about planning, goal setting, deliberate decisions, and focused communication. Our final word belongs to being fluid around those events which cannot be planned, taking advantage of coincidence, and turning crisis into opportunity. In other words, learn how to embrace serendipity.

24.8 Further Reading

[1] R. Batty, "A Mother's Story", Harper Collins, 2015.
[2] S.R. Covey, "The 7 Habits of Highly Successful People - Powerful Lessons in Personal Change", Simon & Schuster, New York 1990.

Summary of Part IV

In this, the final part of the book, we have explored the issues surrounding a maturing career. In this phase, the cycle of success remains a useful way to think about your career but now the elements of position, reputation, skills and leadership necessarily evolve. We encourage our readers to embrace change as an opportunity, be it change driven by external factors, or change that you initiate. We have also revisited the issue of mentoring and have linked it to the important goal of succession planning. Finally, we have closed with thoughts on maintaining a healthy work-life balance and on the importance of having a long-term vision (Fig. 24.2).

Fig. 24.2 "We have closed with thoughts on maintaining a healthy work-life balance and on the importance of having a long-term vision."

This is where our advice ends. The rest is in your hands.

The manufacturer's authorised representative in the EU is Springer Nature Customer Service Centre GmbH, Europaplatz 3, 69115 Heidelberg, Germany. If you have any concerns regarding our products, please contact ProductSafety@springernature.com

Printed and bound by CPI Group (UK) Ltd, Croydon, CR0 4YY

23/03/2026

02076658-0001